Death Waits in the "Dark"

Recent Titles in
Contributions in Military Studies

Creating an American Lake: United States Imperialism and Strategic Security in the Pacific Basin, 1945–1947
Hal M. Friedman

Native vs. Settler: Ethnic Conflict in Israel/Palestine, Northern Ireland, and South Africa
Thomas G. Mitchell

Battling for Bombers: The U.S. Air Force Fights for Its Modern Strategic Aircraft Programs
Frank P. Donnini

The Formative Influences, Theories, and Campaigns of the Archduke Carl of Austria
Lee Eysturlid

Great Captains of Antiquity
Richard A. Gabriel

Doctrine Under Trial: American Artillery Employment in World War I
Mark E. Grotelueschen

Ordered to Die: A History of the Ottoman Army in the First World War
Edward J. Erickson

Surgeons at War: Medical Arrangements for the Treatment of the Sick and Wounded in the British Army during the late 18th and 19th Centuries
Matthew H. Kaufman

Soviet Military Assistance: An Empirical Perspective
William H. Mott IV

The Battle of Ap Bac, Vietnam: They Did Everything but Learn from It
David M. Toczek

The Emperor's Friend: Marshal Jean Lannes
Margaret Scott Chrisawn

An Uncertain Trumpet: The Evolution of U.S. Army Infantry Doctrine, 1919–1941
Kenneth Finlayson

Death Waits in the "Dark"

The Senoi Praaq, Malaysia's Killer Elite

Roy Davis Linville Jumper

Foreword by Charles H. Ley

Contributions in Military Studies, Number 206

GREENWOOD PRESS
Westport, Connecticut • London

Library of Congress Cataloging-in-Publication Data

Jumper, Roy Davis Linville, 1959–
Death waits in the "dark" : the Senoi Praaq, Malaysia's killer elite ; foreword by Charles H. Ley
p. cm.—(Contributions in military studies, ISSN 0883–6884 ; no. 206)
ISBN 0–313–31515–9 (alk. paper)
1. Malaysia. Tentera Darat. Senoi Praaq—History. 2. Malaysia. Tentera Darat—Commando troops. 3. Special forces (Military science)—Malaysia. I. Title. II. Series.
UA853.M35 J85 2001
356′.1673′09595—dc21 00–064051

British Library Cataloguing in Publication Data is available.

Library of Congress Catalog Card Number: 00–064051
ISBN: 0–313–31515–9
ISSN: 0883–6884

First published in 2001

Greenwood Press, 88 Post Road West, Westport, CT 06881
An imprint of Greenwood Publishing Group, Inc.
www.greenwood.com

Printed in the United States of America

The paper used in this book complies with the Permanent Paper Standard issued by the National Information Standards Organization (Z39.48–1984).

10 9 8 7 6 5 4 3 2 1

Copyright Acknowledgments

The author and publisher gratefully acknowledge the following:

All maps are reprinted with permission of Clint Thomas.

All photographs were provided by and reprinted with permission of the Royal Malaysia Police.

Contents

Photo essay follows Part III.

Maps

Abbreviations

AAP — Aboriginal Auxiliary Police.

ARVN — Army of the Republic of Vietnam. The army built up by the United States in South Vietnam.

ASEAN — Association of Southeast Asian Nations.

BBS — Borneo Border Scouts. A primitive collection of indigenous troops, mostly Dayaks from Sarawak, taken over from John P. Cross and developed by R.O.D. Noone into a clone of the Senoi Praaq.

BN — Barisan Nasional. The ruling coalition group comprised of UMNO, the Malaysian Chinese Association (MCA), and the Malaysian Indian Congress (MIC).

BOA — Bamboo Operations Area. That part of Malaya/Malaysia under DOA control.

BOC — Bamboo Operations Committee.

BRIAM — The British Advisory Mission to the Government of South Vietnam. Headed by Sir Robert G. W. Thompson, 1961-1965.

CCO — Clandestine Communist Organization. Active on Borneo during the Indonesian Confrontation. Sometimes called the Communist Terrorist Organization (CTO). Not to be confused with the peninsular communists (MCP).

CIA	Central Intelligence Agency (United States).
CID	Criminal Investigations Division (Malaysia).
CIDG	Civilian Irregular Defense Group. The successor to and amalgamation of a half-dozen or so paramilitary type operations in South Vietnam.
COMINT	United States Communications Intelligence.
CONUS	Continental United States. U.S. military vernacular for the United States land mass between the Pacific and Atlantic Oceans.
CPMAL	Communist Party of Malaysia. The revamped MCP that remained a thorn in GOM's side well into the 1980s.
CRW	Communist Revolutionary Warfare. An acronym employed by the British to differentiate communist sponsored trouble from garden variety disturbances and/or the usual bush wars.
CTs	Communist Terrorists. Coined during the Malayan Emergency as a term of reference for communist combatants inclusive of Min Yuen members.
DOA	Department of Aborigines and parent organization to the Senoi Praaq. Established in 1950 and first led by P.D.R. Williams-Hunt. However, R.O.D. Noone presided over its critical expansionary phase and created the Senoi Praaq.
DRVN	Democratic Republic of Vietnam (North Vietnam).
FARELF	Far East Land Forces (Great Britain).
FULRO	United Front for the Liberation of Oppressed Peoples. A Montagnard based political organization active in South Vietnam's central highlands.
GOC	General Officer Commanding (in British Army terms).
GOE	Government of Egypt. Similar abbreviations are used by United States Department of State personnel to refer to most world governments.

GOF General Operations Force. The new designation put in place by GOM for the PFF. Its motto, "Speed, Efficiency and Courtesy," was chosen for public relations purposes to demonstrate that the RMP is a people-friendly organization. This catchy sounding little phrase when translated into Bahasa sounds much less ominous than some variation of a "Death Before Dishonor" type slogan.

GOI Government of Indonesia.

GOM Government of Malaysia. Malaysia gained her independence from Great Britain in 1957 and has a Westminster Parliamentary style of government. GOM is led by His Excellency, the Prime Minister, Dr. Mahathir bin Mohamed and the United Malay National Organization (UMNO).

GVN Refers in a Senoi Praaq context to the Government of South Vietnam.

IGP Inspector General of Police. The titular head of the RMP. Directs law enforcement in Malaysia, controls substantial paramilitary forces.

JHEOA Jabatan Hal Ehwal Orang Asli. The Department of Orang Asli Affairs for peninsular Malaysia. Originally known as the Department of Aborigines (DOA) and initially created during the Malayan Emergency to work with the Orang Asli and wean them away from MCP influence.

JWS Jungle Warfare School. Members of the Senoi Praaq imparted their jungle survival/tracking skills to Commonwealth and other forces at this institute located near the village of Kota Tinggi. One of several military type training facilities set up by Great Britain and Malaya to foster their own activities and later to transfer their knowledge to the U.S. and U.S. sponsored forces fighting in Vietnam.

KKO Korps Komando Operasi (Indonesian Marine Commandos). Active during the landings at Pontain and elsewhere along Malaysia's east coast. They were excellent troops and were well respected by British and Malay personnel.

KMT Kuomintang. The official ruling party of the Republic of China (i.e., Taiwan) and designation for anti-communist Chinese.

LLDB Luc Lang Duc Biet. South Vietnam's special forces. President Diem attempted to direct these troops himself and initially at least they provided him with leverage over conventional military forces.

MAAG — Military Assistance and Advisory Group. This command was setup in 1955 and managed the US military effort in Vietnam prior to escalation and formation of MACV.

MACV — Military Assistance Command, Vietnam. Formed in 1962. MACV absorbed MAAG two years later in 1964.

MAF — Malaysian Armed Forces. United States designation for the Malaysian military.

MCP — Malayan Communist Party. The political organization led by a Malayan Chinese, Chen Peng, sponsored to a limited extent by the Peoples Republic of China, which offered ideological legitimacy to the insurgents.

MIS — Ministry of Internal Security. Controlled the Senoi Praaq at one time but is now defunct.

MOD — Ministry of Defense. Refers in the context of this work to the Malaysian Ministry of Defense.

MNUSD — Ministry of National Unity and Social Development. This body formerly oversaw the activities of the JHEOA.

MPABA — Malayan Peoples Anti-British Army. Successor to the MPAJA and forerunner to the MRLA.

MPAJA — Malayan Peoples Anti-Japanese Army.

MRLA — Malayan Races Liberation Army. The military arm of the MCP.

NKNA — North Kalimantan National Army. The front organization used by Indonesia to strike the first blows during the Indonesian Confrontation (on Borneo).

NVA — North Vietnamese Army.

OSCAJA — Overseas Chinese Anti-Japanese Army. Military arm of the KMT eradicated by communist Chinese forces prior to commencement of hostilities against the governments of Malaya and Great Britain during the Emergency.

PAAC Perak Aboriginal Areas Constabulary.

PAG Police Aboriginal Guard.

PAS Parti Islam Se-Malaysia. Malaysia's Islamic fundamentalist political party and constant thorn in UMNO's side.

PFF Police Field Force. The counterpart to regular military forces that worked very closely with civilian intelligence organizations both in Malaya and Vietnam. Very useful in this respect but also quite practical in the sense that such obfuscation helped to deflect hostility away from the regular military.

PG Persian Gulf. The wellspring of cheap oil and *sine qua non* of U.S. Middle East Policy. A large part of the cost of this oil is associated with its delivery to market. Any deviation from existing supply routes, many of which pass through the Straits of Malacca, would drastically raise its price. The economies of key United States trade partners, Japan, Korea and Taiwan, would be badly affected.

PGA Police Gerakan Am. The Bahasa designation for the Police Field Force. Currently no longer in use.

PKI Partai Komunis Indonesia. Indonesia Communist Party.

POASM Persatuan Orang Asli Semenanjung Malaysia. A national Orang Asli association with branches in most states dedicated to advancing the Orang Asli position in Malaysian society. POASM operates as a proto political party and is the flagship organization of the Orang Asli political movement. POASM was initially sponsored by the JHEOA and the two organizations were at first on very friendly terms. However, POASM began its break away from JHEOA tutelage under the leadership of Majid Suhut.

PRC Peoples Republic of China.

RAF Royal Air Force (i.e., that of Great Britain).

RMP Royal Malaysia Police. One of the more important institutions bequeathed by Great Britain to Malaya upon Independence. The tradition of excellence instilled in the RMP by Great Britain is comparable to that of police forces in Canada and Australia.

RMR Royal Malay Regiment. That aspect of the MAF dependent exclusively upon Bumiputra sources of manpower.

RPKAD Indonesian Para Commandos. Elite troops active on Borneo and later dropped at Labis by the Indonesians.

RTG Royal Thai Government.

RTP Royal Thai Police.

RVN Republic of Vietnam (i.e., South Vietnam).

SAS Special Air Service. The elite British force that trained the Senoi Praaq in the use of weaponry and assault tactics. The SAS also gave the Senoi Praaq the right to wear the SAS red beret. However, since 1958 all SAS units have worn beige and not red berets, in keeping with wartime SAS practice. Note further that Australia, New Zealand and the United States also maintain their own versions of the SAS.

SBS Sabah Border Scouts. Set up/taken over by R.O.D. Noone in May 1964 using mainly Murut tribesmen and headquartered at Keningau in Sabah. These forces were then operated in conjunction with BBS elements heretofore led by John P. Cross. Subsequent to R.O.D. Noone's departure for Thailand the SBS came under the control of Charles H. Ley.

SEATO South East Asia Treaty Organization. Originally established as part of an alliance strategy to help mobilize support for American policy in Southeast Asia. Unlike NATO, SEATO is now defunct. SEATO did not fair well due to SVN's collapse.

SEP Surrendered Enemy Personnel. During the Malayan Emergency SEPs were treated much better by the government than MRLA members taken captive subsequent to engagement with security force personnel. Many SEPs were recycled into the struggle against the communists.

SOVF Special Operations Volunteer Force. Usually comprised of SEPs. However, some of the early SOVF like units also recruited Orang Asli.

SVN South Vietnam.

TNM Tentara Nasional Malaya. A front organization similar to the NKNA put into place by GOI during the peninsular phase of the Indonesian Confrontation. Specifically, the TNM purported to be the army of the Malayan people seeking liberation from neo-colonial domination.

UMNO United Malay National Organization. The political party which controls the Barisan Nasional (BN), a coalition of several ethnic based political parties, and which has dominated Malaysian politics since 1957.

USAID United States Agency for International Development. A branch of the United States Department of State that dispenses foreign aid and thus facilitates the cultivation of ties with third world leaders.

USOM United States Operations Mission. The designation of the United States aid agency operating in South Vietnam in the 1950s.

Foreword

Death Waits in the "Dark" brings to light with great clarity the legend of heroic exploits by Malaysia's unique paramilitary fighting force, the Senoi Praaq. This remarkable study takes the reader into peninsular Malaysia's tortuous jungle where Senoi Praaq troopers successfully engaged communist terrorists in protracted, yet defining, battle for control of the newly independent state; then to the war in central Vietnam where a contingent of Senoi Praaq troopers and British officers sought to build a fire wall among indigenous mountain tribes in a secretive, failed attempt to shut down the progression of Viet Cong fighters into South Vietnam's southern and eastern provinces; and finally, to the island of Borneo and to the southmost states of Malaysia's mainland where the troopers helped play a signal role in turning back invading forces from Indonesia.

I was privileged to work with the enigmatic Dick Noone during these events which Roy Davis Linville Jumper so ably brings to life and I have now had the pleasure of reliving some very interesting days. He is to be congratulated on his noteworthy achievement in drawing together a morass of details and opinions concerning a chain of activities when few reliable records were ever made—particularly so inasmuch as inevitably opinions and the degree of veracity of accounts vary somewhat from source to source in these matters. My own recollections and thoughts are no exception.

This first study of Malaysia's unique fighting force fills a significant void in the military history of Malaysia and Southeast Asia. It also provides some hardy grist for the mill of readers who are concerned today with Malaysia's national security issues and the stability of the region along the Malacca Straits and the South China Sea.

Charles H. Ley
Wiltshire, 2001

Preface

This study provides the reader with a straight forward, realistic account of Malaysia's legendary special forces unit, the Senoi Praaq. The unit, originally a creature of British military intelligence, was set up in 1956 to fight communism during the Malayan Emergency. Its troopers fought in the deep jungle as Special Air Service (SAS) proteges, eventually replacing Her Majesty's best upon Malayan independence from Great Britain. They then served as mercenaries employed under South East Asia Treaty Organization (SEATO)/United States Central Intelligence Agency (USCIA) auspices in Vietnam and a short time later fought on Borneo during Malaysia's own undeclared war with Indonesia.

Today the unit, largely comprised of non-Malay tribal peoples known collectively as the Orang Asli of peninsular Malaysia, remains under arms and heads up a large paramilitary apparatus maintained in conjunction with conventional military forces. I contend in this research that the Senoi Praaq played a pivotal role in Malaya's transition from colony to independent nation and remains today a unique fighting force upon whom Malaysia must rely to preserve her national security interests both within Malaysia proper and along the perimeter of the vital international passage route, the Malacca Straits.

The evidence and argument put forward in my analysis drawn in important measure from interviews with veteran Senoi Praaq commanders and troopers refutes the contentions of social activist researchers who hijacked the persona of the Orang Asli tribes in the 1960s and continue to this day to depict Malaysia's indigenous peoples as primitive, lost and helpless victims of modernization who are disgruntled wards of the state. Their publications harm the very peoples they purport to champion. Nonetheless, beefed up political pamphlets are hustled towards just such an end thinly disguised as academic scholarship and published by a variety of institutes and centers for this or that Orang Asli concern. The authors of these works only replicate in Malaysia a pattern spun by advocacy oriented writers in other countries around the world, *vis-à-vis* a multitude of indigenous peoples, but hardly applicable in Malaysia. A discerning observer understands however that the Orang Asli peoples—albeit a disadvantaged ethnic group—constitute an absolutely indispensable human resource which needs to be cultivated and channeled over time into productive rather than confrontational relations with the state.

While my study pays the Orang Asli their due as a people it also shatters the passive and stereotypical stigma which marks their public image. *Death Waits in the "Dark"* is not a wholesome book about nice people, friendly places and quaint things. Ambushes, assassinations and assorted dirty deeds, the very essence of irregular warfare in counterinsurgency operations, are taken to chilling extremes by the unit's Orang Asli, Malay and British personnel. Their story is a fast paced and often graphic account of a virtual killing machine in action. Malaysia, Thailand, Vietnam, Borneo and Singapore are all venues of Senoi Praaq operations where the business end of the intelligence game is brought to bear and played to the max.

If the Orang Asli embody the altruistic qualities often associated with Jean Jacques Rousseau's much heralded noble savage they likewise exhibit any number of dark Hobsian traits conveniently ignored by some researchers. This ugly truth points up admirably the fact that there is no anthropological or sociological Shangri-la wherein resident peoples are incapable of being taken and trained either to kill or to kill more efficiently. Killing for that matter comes easy in the jungle.

The evidence assembled in this research corroborates the paradigm in Orang Asli studies first put forward by John D. Leary, who brought us *Violence and the Dream People,* and extended by myself in several studies of the Orang Asli polity. These texts make one point perfectly clear: The Orang Asli peoples give as good as they get. They are not stereotypical patsies. Time and time again they prove themselves able to achieve satisfaction in return for their troubles whether they be political or military.

Many people were helpful in the preparation of this book. Some do not wish to be acknowledged in the usual manner since they remain active professionally and are involved in sensitive matters. In these instances a generic thank you to the Royal Malaysia Police is in order. However, several individuals now retired or otherwise employed may be thanked for their most helpful contributions. These men are of different nationalities and come from a variety of backgrounds but they all share an abiding interest in the Orang Asli. They deserve a special "thank you" and include Mohamed Ruslan bin Iskandar Abdullah, Sambut Anak, Haji Ahmad bin Khamis, Mike G. Jones, Charles H. Ley, Desmund Lawrence, and Mat Noor.

Finally, I am particularly indebted to Clint Thomas, Professor Emeritus, Eastern Michigan University and formerly Professor Indiana University Shah Alam, Malaysia who once again constructed the necessary maps and helped in countless other ways. This is the third book he has helped me pull together and there really is no way to adequately acknowledge such support in words but I can try—thank you Clint.

Roy Davis Linville Jumper
Kuala Lumpur 2001

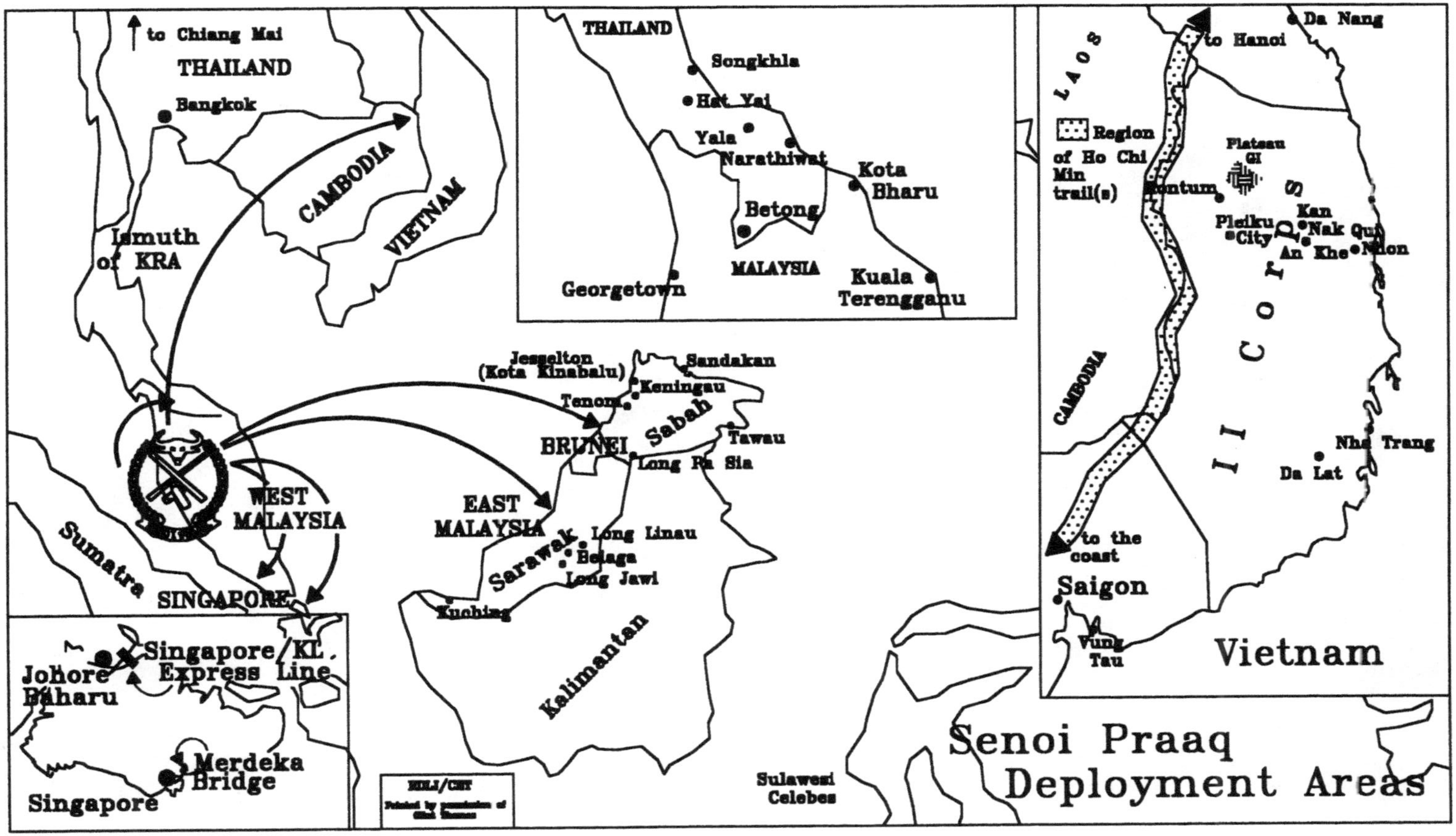

Senoi Praaq
Deployment Areas
to Chiang Mai
THAILAND
Bangkok
CAMBODIA
VIETNAM
Ismuth of KRA
WEST MALAYSIA
Sumatra
SINGAPORE
Singapore/KL Express Line
Johore Baharu
Merdeka Bridge
Singapore
THAILAND
Songkhla
Hat Yai
Yala
Narathiwat
Kota Bharu
Betong
MALAYSIA
Georgetown
Kuala Terengganu
Jesselton (Kota Kinabalu)
Sandakan
Keningau
Tenom
Sabah
Tawau
BRUNEI
Long Pa Sia
EAST MALAYSIA
Sarawak
Long Linau
Belaga
Long Jawi
Kuching
Kalimantan
Sulawesi Celebes
LAOS
Region of Ho Chi Min trail(s)
to Hanoi
Da Nang
Plateau GI
Kontum
Pleiku City
Kan
Nak Qui
An Khe
Nhon
II Corps
CAMBODIA
Nha Trang
Da Lat
to the coast
Saigon
Vung Tau
Vietnam

Death Waits in the "Dark"

Chapter 1

Introduction

One of the world's finest jungle fighting forces—the Senoi Praaq—is the focus of this study. Senoi Praaq fame as Malaysia's killer elite can be traced to the Malayan Emergency and to its legendary performance in the protracted struggle against communist revolutionary warfare in the vast dark jungle interior beyond the northern rim of the Malacca Straits. These men recruited from the Orang Asli tribes proved to be breathtakingly successful in the suppression of the Malayan Races Liberation Army (MRLA), so much so that at one point in the conflict more communists met their fate from Senoi Praaq hands than from any other security forces unit.

These prize troopers remain under arms today and form an integral part of Malaysia's national security apparatus. They continue to enjoy an exclusive mandate for intelligence gathering among the Orang Asli and are subject to Home Ministry control for this purpose. The unit's unique role thus compares to none other within the greater Malaysian defense establishment.

STORMY WATERS

Bordering Malaysia the Malacca Straits has long been associated with great wealth, trade, intrigue and war. The history of this region which brackets the lucrative trade route between Europe and Asia is much discussed, very complicated and difficult to decipher.[1] However, this simple outline clues us in on the essential story: fabled wealth begot trade and migration while intrigue soon followed as did war.

Several different ethnic groups gravitated to the Malacca Straits area in search of its rich resources. These peoples staked out territory and their kingdoms rose and fell as turmoil reigned in what would only much later become generally known as the Malay archipelago. Strife and convoluted realism were the norms. Stability on the Malay peninsula itself remained elusive for centuries. For our purposes the terminus to this tumultuous sequence of events began some five hundred years ago when two indigenous ethnic groups—Orang Asli[2] and Malay[3] —evolved parallel and sometimes conflicting leadership structures.[4] Four aboriginal states—Sungi, Ujong, Klang and Jelebu—in what is present day Negri Sembilan are known to have existed. These tribal confederations lost their autonomy over a 150-year period to the Malacca and Johore Sultanates.

Malay Sultans and Sakai[5] Rajahs sought mutual accommodation with one another as they struggled to survive successive waves of Portugese, Bugis and Dutch encroachment. However, these alliances eventually fell apart and the peninsula lapsed into a period of anarchy.[6] The instability engendered by this sorry state of affairs ultimately afforded Great Britain an opportunity to usurp all other regional and European influences and subjugate Malay and Orang Asli alike.

Great Britain briefly shared her pleasures so to speak with Japan during the Second World War but following the allied defeat of the axis powers fate thrust independence upon Malaya in 1957. Then in 1963—in a further act of voluntary unilateral decolonization—Great Britain consolidated the Borneo states and Singapore into a federation with Malaya proper, to form Malaysia. This key albeit circuitous development finally lent a contemporary name and face to a place long known informally as the land of the *Keris*, arguably the most sanguine *arme de main* in the world.[7]

A NATION MATURES

Malaysia per se has not had an easy time since independence and maintenance of her statehood proved to be an expensive proposition. The country defended herself against several attacks by Indonesia in the mid-1960s, experienced aggravated race riots between ethnic Malays and immigrant Chinese shortly thereafter and engaged in a protracted, albeit successful, struggle against communist guerillas well into the late 1980s. The nasty communist insurgency known locally as the Emergency dwarfed Malaysia's ethnic problems not to mention her transitory difficulties with Indonesia.

The sanguine nature of the Emergency very nearly spoiled centuries of difficult labor and could easily have killed off the nascent state. However, the country is stable now and since those dark days Malaysia's problems have shifted from military matters to economic concerns. Several world class recessions, from which the country continues to suffer, have kept real pressure on the Malaysian government.

Thus, neither modernity in general nor independence in particular were painless processes. They did not come cheaply. However, there is an upside to Malaysia's trials and tribulations which more than offsets her ordeal. The country benefits from these past difficulties in the sense that her leadership has been tested and is by this point extremely savvy.

The Government of Malaysia (GOM) enjoys an advantage in sophistication acquired through many hard-learned lessons. It is alert on this account to any number of ruses liable to ensnare or topple a young regime. For that matter

Malaysia the "state" may be young in a formal legal sense but its substantive underpinning is more mature and much wiser than initially meets the eye of many observers from the outside.

THE MALAYSIAN WORLD VIEW

GOM policies are thus laden with an acute and well-deserved sense of paranoia over perceived threats to Malaysian sovereignty. A national consciousness simply does not evolve over centuries burdened with protracted civil wars, colonization, occupation and a communist inspired attempt at revolution without picking up a few scars. The anxiety level in the Malaysian political psyche is in this regard fairly high but it is well within the rational actor range and on par with that exhibited by several other nations in the Far East. South Korean apprehension of the Communist north comes to mind as does the traditional Vietnamese fear of China and the general distaste of Asians for the Japanese.

The protectiveness of her sovereignty Malaysia exhibits is low key, purely defensive in character and not focused upon any one particular country. Pragmatism born of hard experience accounts for GOM's independent mindedness and adoption of a quasi-non-aligned stance on the world stage. This posture which affords a small country some degree of autonomy has in fact served Malaysia well in the past. However, Malaysia's past tribulations pale in comparison with the more ominous threats which loom large on the horizon. A brief review of these regional and international issues provides an appropriate frame of reference for our study of the Senoi Praaq, a precious jewel in Malaysia's national security arrangements.

THE MALACCA STRAITS: AND A NEW PATTERN OF POWER

Several new perils invest of late the thoughts of Malaysia's foreign policy planners. Chronic economic problems and ethnic woes make Indonesia an unstable neighbor; Malaysia's claim to the Spratley Islands may well be challenged by China; and Singapore, once an integral part of the Malaysian Federation, is armed to the teeth and ever mindful of Malaysia's own restless ethnic Chinese population.[8] However, some GOM policymakers are more concerned with the dangers emanating from the projection of United States military power into the more immediate region around the Malacca Straits than with these disputes with Malaysia's neighbors.

GOM views this particular development as a natural by-product of the Cold War's end, United States loss of permanent base facilities at Subic Bay in the Philippines and America's heightened dependence upon trade with Asia. There is essentially no real opposition to United States hegemony in Asia and to its assertive steps to gain access to significant staging facilities outside of Guam in order to guarantee continued stability in the region.

For that matter the assertive military strategy arrived at to compensate for loss

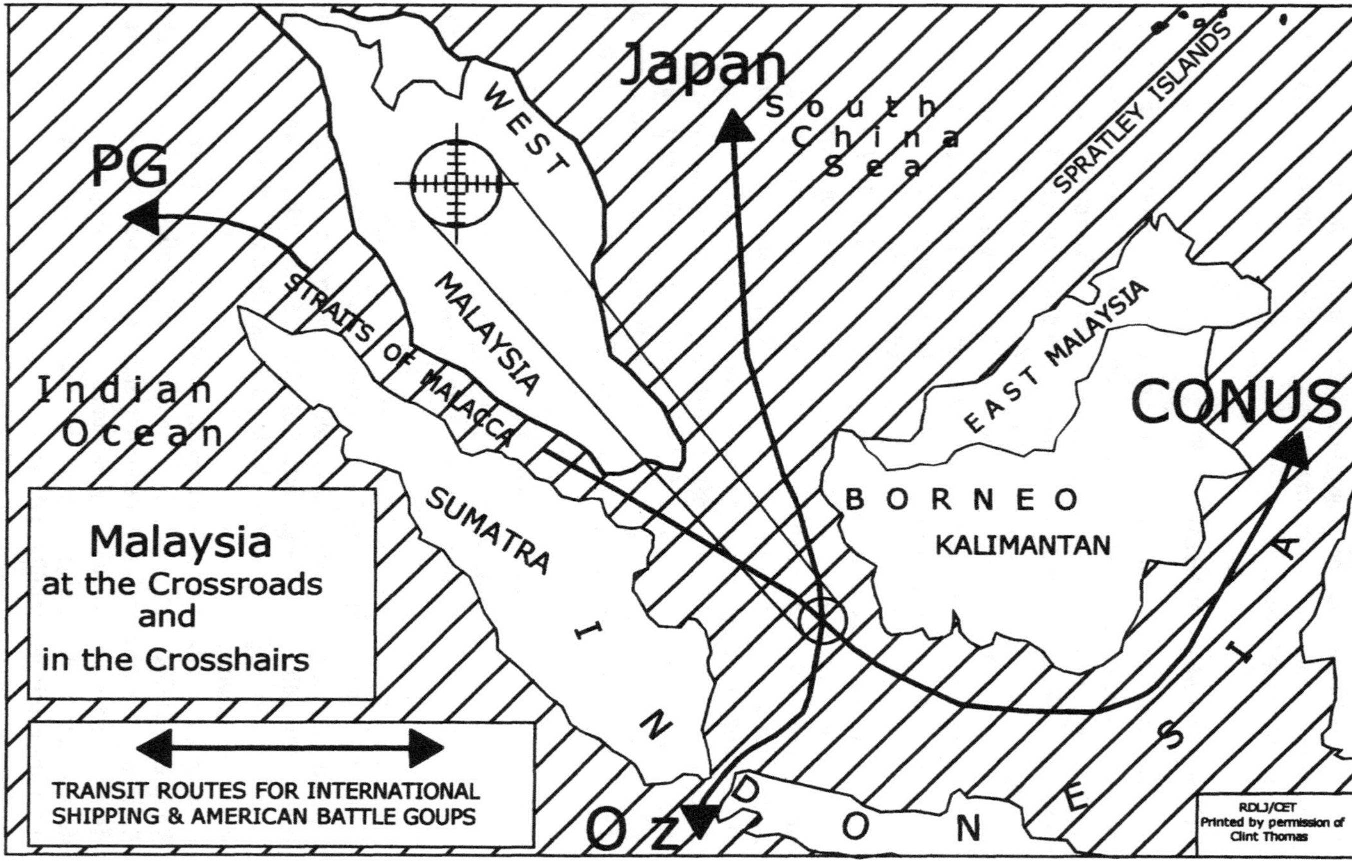
Japan
South China Sea
SPRATLEY ISLANDS
PG
WEST MALAYSIA
STRAITS OF MALACCA
Indian Ocean
SUMATRA
EAST MALAYSIA
BORNEO
KALIMANTAN
CONUS
INDONESIA
Oz
Malaysia
at the Crossroads
and
in the Crosshairs
TRANSIT ROUTES FOR INTERNATIONAL
SHIPPING & AMERICAN BATTLE GOUPS
RDLJ/CET
Printed by permission of
Clint Thomas

of the Philippines as the United States center of gravity in the western Pacific entails general litoral engagement whereby American naval forces transit the region cyclically on north/south and east/west axes. Instead of massing around the Philippines these carrier battle groups now shuttle back and forth between the Continental United States (CONUS) and the Persian Gulf (PG) in a lateral direction and vertically from Japan down to Australia and back. This manoeuver in which the United States conducts regular joint-exercises with various local forces has become known in Navy vernacular as the Southern Swing.

Such massive asset movement makes substantial air power available for commitment within a week to ten days at any point in the region. It also dictates for logistical reasons that numerous courtesy ports be available to resupply the ships and the people involved. This in turn necessitates the pursuit of cooperative engagement and various alliance type arrangements with host governments well outside of East Asia proper.

GOM'S REJOINDER

GOM perceives the United States commitment to globalism, cheap Middle Eastern oil and affordable goods from Asia as absolute and it recognizes for this trade management strategy to work the United States must take the steps necessary to keep the Malacca Straits open. At a minimum Malaysia and Indonesia have to be suborned or at least made tame as the basis of any sort of free trade scheme in Asia. Hence, Malaysia sees a web of regional security alliances spun around her and cannot help but correlate these developments to events that took place about twenty years ago at another significant world crossroads.

Simply put Malaysia does not want to become co-opted like Egypt which sits astride a similar choke point in an international sea passage—the Suez canal. Under Gamal Abdel Nasser's regime this country maintained a high degree of autonomy despite extensive ties to the Soviet Union. Since those heady post-Farouk days so filled with promise Egypt has gravitated into a United States orbit where she no longer has the freedom to act in a realist/rationalist manner in the conduct of foreign relations.[9] Gone perhaps forever are her own aspirations for regional superpower status and any possible hope of slipping onerous ties to Israel.

GOM maintains that the Egyptians fell into a dependency relationship because "power" in Egypt never did disappear into the "center"of government as it so often does in representative democracies. Pharonic visibility, pomp, circumstance and improper safeguards made control over the reigns of power vulnerable to export offshore in the wake of an alliance with the United States. It was just too easy for American operatives to get at those Egyptians who straddled key action channels in government and induce their fealty through largess and other forms of persuasion.

The élan with which United States diplomatic personnel fawned over Government of Egypt (GOE) officials could not go unnoticed and was a sight to behold. This was a short-term debasement of sorts, that stands in stark contrast to the more recent military bravado in Iraq, but in the end the desired result was achieved. Egypt is now hooked on foreign aid and as a telltale reminder of this fact boasts possibly the most ostentatious United States Agency for Development (USAID) building in the world.[10] GOM policy planners are determined that this course of events, or some other carrot-and-stick type fate similar to that which befell the Noriega government in Panama, should not befall Malaysia. They are ever wary of United States intentions.

THE MALAY WAY

Such apprehension also explains why Malaysia's political community remains so insular and accounts for much of the hostility towards the United States despite extensive trade ties which exist between the two countries. Regional diplomatic initiatives undertaken by the United States to leverage Singapore as a mini-superpower surrogate aimed at the soft underbelly of the Malay peninsula are a real sore point. In the GOM view Singapore now poses a latent threat to Malaysia just like Israel has been built up as a counter force to several of the Arab states.[11] However, most GOM representatives are equally quick to point out they do differentiate between America's real interests as a hegemonic power and the misguided policies of pro-Israeli diplomats they firmly believe run the show in the United States Department of State.

The bold extension of United States naval power into the region of the Malacca Straits together with the continuing litany of complaints about Malaysia's financial policies and human rights record are matters of some concern to the GOM. Vice President Albert Gore's testy in-your-face speech in Kuala Lumpur reproaching Malaysian officials for the arrest of the sacked Deputy Prime Minister, Anwar Ibrahim, is especially vexing to them. It only exacerbated the already troubled relations between the American Embassy's poorly led and moribund political section and GOM officials.[12] The Mahathir government considers itself to be among the most tolerant in Asia and very nearly the victim of an attempt to subvert the country's leadership element through non-constitutional means. Hence, some Malaysian officials perceive and resent pressure from United States officials made ostensibly in the name of democracy.[13]

United States diplomats who malign GOM and egg on Malaysian dissidents are said to promote a bastardized form of democracy that allows for the easy removal of leaders who do not acquiesce to pressure. American style democracy is viewed in this context as little more than a smoke screen for neo-colonial machinations geared to the mercantile compromise of Malaysia's hard won independence. For that matter all systems can be and often are manipulated to the advantage of their proponents. Democracy is no exception. In a weak pseudo-democratic setting it is a lot easier for hidden interests to manipulate the rules from afar and remove a non-compliant leader from power than it is to send in the Marines.[14]

Nonetheless the GOM, for all its trepidation, continues to play the game so to speak and makes every effort to avoid becoming isolated on the world stage. It conducts limited joint military exercises with a variety of countries inclusive of the United States, purchases weapons packages from multiple sources and forges ahead with defense industrialization.[15] This defense activity is designed to ward off external threats with no intent to take offensive action against neighboring countries. Rogue nation states like Iraq and North Korea are not likely to see Malaysia enroll in their ranks any time soon.

The same spirit of determination and a desire to show their teeth helps to explain the willingness of GOM officials to give the author unusual access to the country's legendary special forces unit, the Orang Asli based Senoi Praaq.[16] This study, reflecting field research in Malaysia's jungle interior, affords the reader a rare glimpse into the Malaysian order of battle and an opportunity to eyeball one of the world's premier jungle fighting forces. The inference to be drawn from the study of the Senoi Praaq is very clear: GOM has both the physical means to preserve its sovereignty and the will to fight to sustain it.

AVENUES OF EXPLORATION AND OBJECTIVES OF THE STUDY

In this book we probe into the unit's origins, analyze its tribal composition, evolution and modus operandi and assess its significance to today's Malaysian defense establishment. We evaluate unit accomplishments, develop a rough unit chronology and research Senoi Praaq performance responding to three critical challenges: the Malayan Emergency which burst forth on the eve of the country's independence, the covert stint of Senoi Praaq troopers in Vietnam and their role later in the Indonesian crisis.

Our objective is to discover the unit's many early accomplishments, to update our knowledge of its recent activities and then to relate these findings to the broader issue of Malaysian national security. The underlying argument presented by this research asserts that the Senoi Praaq is an ideal vehicle for the maintenance of regime stability in Malaysia. Key factors behind the unit's success are identified, discussed in some detail and applied towards substantiation of this premise.

Thus, the general idea is to extend my own earlier analysis of Orang Asli political capacities (outlined in *Power and Politics: The Story of Malaysia's Orang Asli* and *Orang Asli Now: The Orang Asli in the Malaysian Political World)* into the realm of military affairs. As such *Death Waits in the "Dark": The Senoi Praaq, Malaysia's Killer Elite* is the final installment in a trilogy of books about the Orang Asli which bring to light their many, varied and positive contributions to Malaysia.

PART I

The Political Background

The nuances and underlying currents affecting Malaysian security are difficult to grasp without first touching upon some ancillary aspects of Senoi Praaq evolution and development. Multiple relationships among a variety of actors including the Orang Asli tribes come into play but they are easily obscured by events. The Malays, Orang Asli and Malaysia all need to be placed into a rough time, space and action framework wherein some semblance of direction is imparted upon the activities of the Senoi Praaq, the focus of the study. This provides the necessary logic to Senoi Praaq existence and it also serves as a convenient point of departure for analysis of the Senoi Praaq itself. Only then may we isolate the particular role of the Senoi Praaq, conduct an objective study, and be in a position to interpret the results with precision.

In Part I we examine events leading to Malaysia's independence, assess the current state of Orang Asli/Malay relations and affix a Senoi Praaq rubric to the Malaysian order of battle. The Orang Asli in turn are established as an independent variable in Malaysian politics and portrayed as traditional mercenary allies of the Malays, all the while the Senoi Praaq is depicted as GOM's premier jungle fighting force. This analysis lays bare the essential rationale for meticulous inquiry and paves the way for the study of the paramilitary unit itself.

Chapter 2

The Orang Asli and Malay Peoples in Perspective

GOM depends upon non-Malay sources of manpower for national defense and makes extensive use of a heterogenous group of tribal peoples known collectively as the Orang Asli.[17] There are three broad tribal categories subsumed under the Orang Asli label—the Semang, Senoi, and Jakun[18]—each of which possesses its own sub-components.[19] They are intimates of Malaysia's forbidding canopy jungle and long time mercenary allies of the ethnic Malays who control Malaysia today.

Orang Asli origins are obscure[20] but the Malay and Orang Asli relationship itself dates to the early days of Orang Asli dealings with Parameswara, the refugee prince from Sumatra and first great Malay leader who founded the Malacca Sultanate in 1400-1401 A.D.[21] Malacca city, a river mouth port, quickly developed into Southeast Asia's main trading center whose importance compared to none other than perhaps to Alexandria, Egypt. Parameswara invested this site with absolute authority over his subjects along the Malacca Straits. The fact that such an audacious prince, who could have anything he wanted, ultimately chose Malacca—as did the Portugese, and the Dutch—for the location of his headquarters is supreme testimony to the waterway's significance.[22]

Parameswara for his part established a highly successful kingdom, a virtual trading emporium buttressed by Orang Asli military services. Malacca was the model for other principalities who joined together ultimately to form today's Malaysia. Malay power maintenance arrangements on the peninsula more often than not included the Orang Asli in some way. From the time of the Malacca Sultanate onwards the Orang Asli and Malay have collaborated across a spectrum of activities but the Orang Asli military function itself remained the singular constant in this relationship.

Their past experience is best described as one of intermittent cooperation—punctuated by periods of strife—in which the Malays gradually gained the upper hand and subordinated the Orang Asli to their rule. As mercenaries and retainers for various Malay sultans the Orang Asli tribes became a card in someone else's

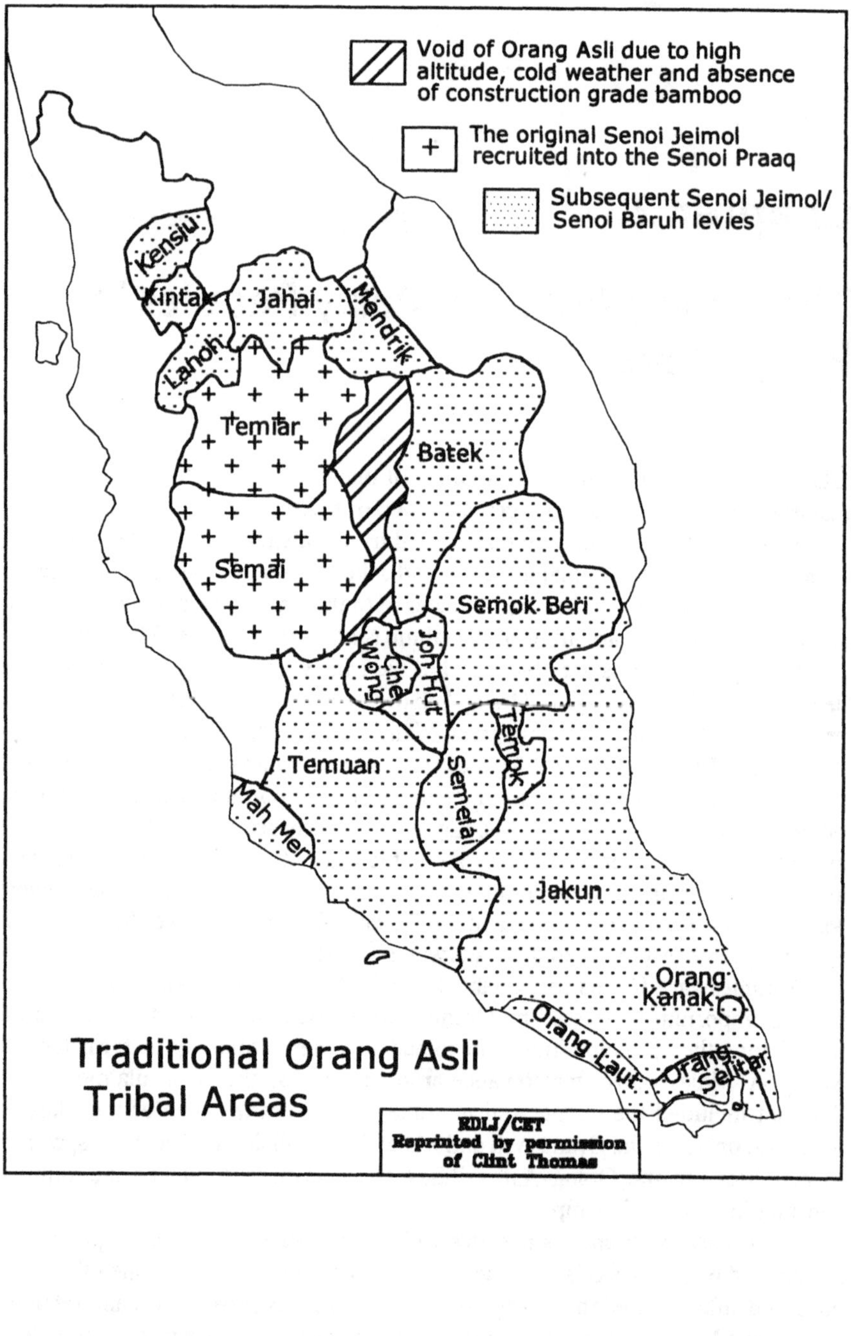
Void of Orang Asli due to high altitude, cold weather and absence of construction grade bamboo
The original Senoi Jeimol recruited into the Senoi Praaq
Subsequent Senoi Jeimol/ Senoi Baruh levies
Kensiu
Kintak
Jahai
Mendrik
Lanoh
Temiar
Batek
Semai
Semok Beri
Che Wong
Jah Hut
Tembok
Temuan
Semelai
Mah Meri
Jakun
Orang Kanak
Orang Laut
Orang Seletar
Traditional Orang Asli Tribal Areas
RDLJ/CET
Reprinted by permission of Clint Thomas

deck but as survivors in a turbulent environment they are no strangers to violence. Indigenous Malay and Orang Asli alike share a tempestuous heritage and still today they are mercurial allies bar none.

TRADITIONAL ORANG ASLI/MALAY RELATIONS AND ECONOMIC DEVELOPMENT

The Orang Asli/Malay relationship itself is quite ambiguous and there is something elusive between these two peoples which is most difficult to describe and pin down. Little similarity exists between their relationship with each other and their respective relationships with other ethnic groups in Malaysia. This is because in times past Malay royalty from Parameswara onwards practiced a policy of intermarriage with the Orang Asli to achieve legitimacy through matrimonial alliances.[23]

Thus, some Orang Asli and Malay groups did intermix over the centuries but the lines along which this melding occurred are rather blurred. It would be improper to lump them into a single racial composite. Many Orang Asli and Malay share few cultural similarities despite prolonged exposure to one another and they are best described as related but different peoples. Consequently, latent tensions lurking beneath a facade of normalcy do permeate the present Orang Asli/Malay relationship despite recent efforts by GOM to integrate the Orang Asli into Malaysian society.

To ignore the friction between the Orang Asli and the Malay would be deceitful, do a disservice to both ethnic groups and belittle GOM efforts to accommodate the Orang Asli. By the same token and to place matters in their proper perspective it is also fair to say this strain in their relations, infrequently manifested in any overt sense, is much more muted than Malay/Chinese antagonism. Let us simply say the Orang Asli and the Malays are deeply entrenched together in a symbiotic bond with an intense emotional dimension.

Scholars often point to contradictions between Islam, the religion of the Malays, and the animistic beliefs of the Orang Asli as a prime source of contention between the two peoples. This traditional explanation of their differences may suit some purposes but in truth Malayan Islam incorporates elements of an Orang Asli cosmology based upon animism. Perhaps this overlay best be viewed as a mitigating circumstance rather than a source of dissension. Orang Asli/Malay strains today can be attributed most likely to material rather than spiritual issues. Development encroaches increasingly upon traditional Orang Asli homelands yet few Orang Asli accrue benefits from this gain in which Malaysia itself becomes a relatively wealthy nation. Consequently, many Orang Asli face a dilemma, either integrate into the broader Malaysian society in the hopes of sharing in its wealth or attempt somehow to continue their detached, semi-nomadic existence as shift and burn agriculturists.

However, the once inexhaustible supply of virgin jungle within which the Orang Asli can retreat and follow the old ways has become a finite commodity.

Bamboo, the primary material the Orang Asli need for shelters and house building, cooking utensils and various other essential tools, ceases to proliferate at about 4,000 feet above sea level and a bamboo culture becomes impossible to sustain beyond this point although the main range rises above 7,000 feet in some places. Many Orang Asli are thus pinned between this "bamboo ceiling" which separates Malaysia's main range from the palm oil plantations, rubber estates, tin mines and industrial projects springing up at lower elevations. The bamboo ceiling as the marker of a potential national security flashpoint underscores the Senoi Praaq's value in sustaining existing political and military arrangements in Malaysia.

NO IDEAL TYPE ORANG ASLI

Quite a few Orang Asli have a Mon-Khmer physical quality resembling the Montagnards found in the highlands of Cambodia, Laos and Vietnam. This is particularly true of the Senoi peoples inhabiting Malaysia's main range. Most anthropologists would probably agree that their tribal groups migrated southwards through Thailand into peninsular Malaysia many thousands of years ago.

However, the Jakun and Semang peoples are two exceptions to this general understanding of Orang Asli roots. The Jakun—often referred to as Proto Malays—live in southern regions of the country and share some common racial and physical characteristics with the Malays but from a cultural perspective they are a distinct and different people. The Semang for their part live in the north where many of them roam back and fourth across the Malaysia/Thailand border. The Semang are an anthropological puzzle since they resemble neither the Malays nor the Thais in their cultural and physical features.

Thus, there is no ideal Orang Asli type either in terms of physical appearance or in their manner of living. Any particular Orang Asli spans a range of profiles and lifestyles—depending upon his or her tribal affiliation and location. Some Orang Asli are geared towards life in the central highlands and continue to pursue a quasi-nomadic lifestyle while others follow a more sedentary existence similar to that of the Malays along coastal areas and in the south of Malaysia.[24] Hence, their appearances cut across tribal lines and are most definitely not an accurate marker of Orang Asli ethnicity. There is no uniform Orang Asli genre. They are a very difficult group to categorize from either a social or ethnic standpoint, so much so that even before the Orang Asli began to practice inter-tribal marriage their variance and lack of inter-tribal homogeneity was commented upon by researchers. The complexity of recognizing an Orang Asli by his or her physical appearance is so great that outside observers often have trouble distinguishing an Orang Asli from a Malay.[25] Few things may be said about the Orang Asli with any great uniformity but most of them are animists, many remain detached from mainstream Malaysian life, and they are not very well integrated into Malaysia's cash economy.

ORANG ASLI TIES WITH THE MAHATHIR REGIME

Despite these differences among the various tribes and endless speculation over the their true relationship with the Malays per se any number of Orang Asli are in the employ of the GOM, the Malacca Sultanate's present incarnation. The sine qua non of governance on the Malay peninsula is the regime of Mahathir bin Mohamed and his allies in the United Malay National Organization (UMNO) and the Barisan National (BN).[26] The Orang Asli for their part serve throughout the national government, they are heavily concentrated in the Malaysian armed forces and are tied into various intelligence agencies. Most importantly they man the famed Senoi Praaq, Malaysia's premier paramilitary force. Their omnipresent "presence" as a pillar of regime support is truly ubiquitous and unquestionable.

Chapter 3

Legacy of the Malayan Emergency

Malaysia's military and paramilitary communities are fairly extensive in number and relations among these different entities are, like those of her peoples, quite complicated. The byzantine qualities and the sheer size of this martial complex are in keeping with the peninsula's long and turbulent history. However, one recent deadly episode colors the state of Malaysia's affairs above all others. The bloody Malayan Emergency. This drawn out conflict began in the late 1940s but it did not end formally until December 1989 when the last insurgents gave up the fight and a treaty was signed by the GOM with the Malayan Communist Party (MCP).[27]

The Emergency, a communist inspired revolution by Malaya's ethnic Chinese population under the banner of Mao Zedong, played out as one of the world's first modern day guerilla wars. Great Britain coined the novel term "Emergency" ("Darurat" in Bahasa) in order to deny the insurgents political legitimacy. This clever appellation helped to criminalize their activities in the minds of Malaysia's inhabitants and contributed to their ultimate defeat on the battlefield.[28]

However, before any semblance of victory could be achieved the fighting swept across the entire Malay Peninsula and deep into the main range, like a brush fire, sparing none of Malaysia's ethnic groups from harm. Only late in the day did GOM play its trump card and manage to come to grips with the problem. Government actions against the communists during terminal stages of the war spearheaded by Orang Asli fighters denied the insurgents free reign in Malaya's jungle interior once they had been driven out of the urban areas, fringe jungle and rolling hills common to the southern regions of the country.

The role of the Orang Alsi during the Malayan Emergency invites comparison with the role of the Indochinese Montagnards who functioned as auxiliary forces of the United States during the Vietnam conflict. The two wars overlapped and British experiences in Malaya were a precursor to American entry into the struggle against the communist insurgency in Vietnam. But unlike Vietnam, the Malayan Emergency was finally contained due in no small part to effective indigenous participation in the conflict. United States efforts through South East Asia Treaty

Organization (SEATO) auspices to apply in Vietnam the British/Senoi Praaq knowledge from Malaysia proved to be abortive.

THE MALAYSIAN DEFENSE ESTABLISHMENT

The Emergency traumatized Malaysia and left the fledgling nation state a large defense apparatus that belies the country's small size. Size aside her defense requirements are distinctive in several ways. The establishment's force structure weighs paramilitary capabilities on par with those of regular military assets and its composition is inclusive of non-Malay tribal peoples in a significant capacity.

British policy makers, ever aware of the watchful eye of a parliament loath to wage a full blown war in the defense of Malaya, made unusual use of police personnel and paramilitary forces in order to avoid an overly militaristic solution to the Emergency. This rebellion, in their view, did not differ much from earlier attempts by immigrant Chinese to rise up, usurp Malay prerogatives and overthrow British rule except in its scale and in the introduction of communist ideology into the mix. Thus, this calculated and in retrospect very clever decision to pin the moniker "Emergency" on the rebellion not only denied the insurgents legitimacy, it also deprived them of a distinct focus upon which to channel societal discontent.

Hence, Malaysian paramilitary police and regular military forces today complement and offset one another in an elaborate balance of power arrangement. The Royal Malaysia Police (RMP) for its part has seventeen of Voltaire's big battalions of heavily armed and highly trained troops available for operations in peninsular Malaysia and in Sabah and Sarawak.[29] Most formations have either air or armored assets and some operate autonomously. They recruit mostly from the Bumiputra ethnic category but do not restrict themselves to Muslim Malays as such. The term Bumiputra (sons of the soil) is the politically correct designation for Malay and Malay-like peoples, especially the Orang Asli.

THE SENOI PRAAQ IN THE MALAYSIAN ORDER OF BATTLE

The centrality of the Senoi Praaq, by definition a deep jungle special forces operation, in the conduct of Malaysian political and military affairs cannot be over emphasized. It is one of two elite formations on the civil side of GOM's defense establishment operating at the apex of the RMP's table of organization.[30] The other is presently more of an urban assault type unit (embodied by a special task force) with an equally illustrious jungle fighting past called Commando 69.[31] The two organizations trace their origins to the Emergency. Both units employ a majority of Orang Asli and each is distinguished by great ferocity. Together they form the core of the General Operations Force(GOF).

However the Senoi Praaq, the more traditional and better of the two, is essentially an all-Orang Asli organization commanded by a few Malays. Its domestic repute puts it in the league with the British Gurkhas and the French Foreign Legion and sets it apart from most of the world's fighting units. Yet, the organization's novelty does not stop with these more conventional comparisons as its troopers were trained and commissioned by Great Britain's famed 22nd Special Air Service Regiment (22 SAS).[32]

From an overall mission standpoint in the context of jungle operations Malaysia's Senoi Praaq resembles the SAS itself.[33] However, unlike the SAS the Senoi Praaq operates exclusively in a jungle venue and does not dabble in other areas of irregular warfare. In sum its warriors were the determining factor ending the Emergency in GOM's favor, and they have become a legend in Malaysia. Today, they function as an essential component in the Malaysian government's efficient military and intelligence machine.

Chapter 4

The Senoi Praaq: A Legend Unfolds

Thus, amidst a myriad of rich research opportunities in contemporary Malaysia this book focuses on the Senoi Praaq, the Orang Asli based irregular force British Colonial authorities put on foot in 1956 to fight the communists during the Malayan Emergency. These jungle fighters, a virtual killer elite, were by definition creatures of British intelligence and initially they served as SAS auxiliaries.

Their operations proved to be extremely successful in the suppression of communist combatants recruited for the most part from among ethnic Chinese inhabitants who were principally organized into the MRLA. The Senoi Praaq troopers' unusual success in battle stood them apart from other security forces and at the crucial point in the conflict more communists came to a violent end at their hands than from any other security forces unit. By 1959 their kill ratio stood at 16 to 1 per squadron and peaked higher than that of any other group inclusive of the SAS and the Gurkhas.

Consequently, Senoi Praaq warriors became notorious for their exploits and quickly acquired their reputation as Malaysia's killer elite. The Senoi Praaq came to be the one unit that the communists feared most and dared not to engage even under circumstances advantageous to themselves.

The unit's very name reflects this ferocity. Senoi Praaq, a term taken from the Semai language, means "war people" when translated directly into English. However, the name also reflects extensive collaboration between the Orang Asli and the Malay in the sense that "Senoi" is purely a Semai word meaning "human being" while "Praaq" is a Semai corruption of the Malay term "Perang" meaning war. This corrupted term is pronounced as Praak in English and was actually spelled Pra'ak by the unit's founder although Malaysian government documents no longer incorporate the apostrophe.[34]

Today the Senoi Praaq, the pride of the RMP, monopolize regime oversight across the Orang Asli population-at-large. As such they stand at the center of an obscure and dangerous paramilitary community entrusted with a critical intelligence function. The management of Orang Asli affairs is of such utmost importance to Malaysia's national security that this organization operates under the direct authority of the Home Ministry and its commanding officers are appointed and subject to the oversight of the Prime Minister.

The Senoi Praaq, the only remaining repository of truly significant coercive indigenous power in Malaysia, is one of only three Orang Asli based organizations capable of exerting some degree of influence over the Malaysian political system. The others are the Department of Orang Asli Affairs (JHEOA)[35], which functions as an administrative interface between the Orang Asli and the Malaysian government, and the Peninsular Orang Asli Association of Malaysia (POASM)[36], the premier Orang Asli political organization in Malaysia.

SENOI PRAAQ ORGANIZATIONAL DEVELOPMENT

The jungle areas along the main range have shielded the Orang Asli from Malay encroachment over the ages but they have also been the harbinger of hostile forces who contest the central power of Kuala Lumpur. The Malays could not help but notice that during the Emergency the Orang Asli were indeed mercurial allies who first sided with the communists before the British won them over at great expense. British intelligence operatives engineered an extensive program to deal with this problem and a bureaucratic entity known as the Department of Aborigines (DOA) soon grew up around the Orang Asli. This imaginative creation, initially an intelligence agency thinly disguised as a welfare organization devoted to protection of Orang Asli interests, turned out to be the decisive factor in determining Orang Asli/Malay relations for some years to follow. It provided the Orang Asli their first institutional linkage to the newly independent Malaysian government and, most importantly, it gave the various Orang Asli tribes a common theme around which to unite as one people.

The Senoi Praaq thus evolved as the military arm of a bureaucratic apparatus designed to extract intelligence among tribal peoples of the hinterlands in return for the distribution of material subsidies.[37] However, the Senoi Praaq was soon unleashed from bureaucratic control and for all intents and purposes it functioned as a private army and operated as a mercenary force subject to little oversight. This loose organizational arrangement worked well during the war but for some GOM officials it was no longer appropriate once the communist threat began to subside. Clearly, adjustments were in order and at the earliest convenient opportunity Iskandar Carey—the first post Noone/Herbolt DOA director—oversaw the detachment of the Senoi Praaq from the DOA and placed it under the more immediate supervision of the Ministry of Internal Security (MIS). As a result the Senoi Praaq functioned more or less as an independent entity beginning in 1960. It made its way ultimately to the RMP on February 8, 1968 following a short stint in the now defunct MIS and a further period of direct and less relaxed Home Ministry supervision as a semi- autonomous organization. This reorganization soon after Malaysian independence may be construed as a precautionary step taken by the new governing authority to forestall any problems that might arise in the future from an excessive concentration of Orang Asli power in a single department.

Chapter 5

Orang Asli Auxiliary Organizations Today

There are any number of more diminutive Orang Asli based organizations in existence today and their mention here brings out the linkage between the Senoi Praaq and earlier inquiry into the Orang Asli polity.[38] The development of these organizations reflecting Orang Asli integration into mainstream Malaysian life can be traced directly to the Senoi Praaq. Two of them standout in particular as prototypical entities and serve as model examples of Orang Asli initiative. They are, as mentioned, the Peninsular Orang Asli Association of Malaysia (POASM) and its parent organization, the Department of Orang Asli Affairs (JHEOA).

DEPARTMENT OF ORANG ASLI AFFAIRS (JHEOA)

GOM could ill afford to leave the door to trouble ajar once the Emergency had been contained. The communists would have to be kept at bay for a long time to come and the GOM needed continued Orang Asli support in this endeavor. So, to finalize tastefully the bureaucratic restructuring which sought to move the Senoi Praaq from harm's way by making it accountable to conventional police authority, the GOM then embarked the DOA itself on a long and arduous bureaucratic journey, turning it into the Department of Orang Asli Affairs (JHEOA). This new department which included a substantial Malay administrative element was eventually detached from the Home Ministry and assigned first to the Rural Development Ministry and later to the Ministry of National Unity and Social Development.[39] These bureaucratic maneuvers effectively de-linked the new JHEOA from any intelligence functions and made both the revamped DOA (now the JHEOA) and the Senoi Praaq palatable to GOM.[40]

PENINSULAR ORANG ASLI ASSOCIATION OF MALAYSIA (POASM)

GOM, however, was careful not to deprive the Orang Asli of their political rights altogether and left a safety valve in place. The preservation of Malay prerogatives in an independent Malaysia did not preclude Orang Asli participation in politics as "civilians" once they were detached from the nebulous world of clandestine operations. As a result, a sanitized JHEOA exorcized of armed elements and relieved of its intelligence function not only gave the Orang Asli the necessary sponsorship it also provided them a base from which to launch their own association.[41] Men who retired from the Senoi Praaq were free to take up employment in the JHEOA without stigma once this organization was put on a civilian footing. These operatives recycled as JHEOA staff, who had acquired sufficient exposure through their stint in the Senoi Praaq to aspire to better things in life, formed the nucleus of an association of Orang Asli employees which has evolved into the Orang Asli's most prestigious national political organization—POASM. POASM, officially registered in 1976, held its first annual general assembly in 1977. This event signaled the beginning of Orang Asli political action in independent Malaysia.

Our analysis of the circumstances surrounding the creation of POASM helps assess the Senoi Praaq impact on Orang Asli political development and more importantly on the performance of the Malaysian political system itself. Many of POASM's founding members were indeed the first veterans of the Senoi Praaq. Recruited directly from their tribal homelands without much acclimatization to society at large they were wary of association with other Malaysians. Naturally, the Malays knew these particular Orang Asli for their fierce fighting and sometimes rowdy social behavior. To put it bluntly, the initial group of Orang Asli to decommission after the worst of the fighting in the Emergency had passed and to move into civilian life were a formidable group.

The government's tolerance of their political activities was laden with implications for the future role of this group in Malaysian society.[42] POASM, positioned as a proto-political party ready to get into the rough and tumble of mainstream Malaysian politics, seeks an alignment with the United Malay National Organization (UMNO), the paramount Malay political organization, in which it functions as a satellite party delivering swing votes in state and local elections. POASM expects its success in these elections will help it gain access to the lucrative patronage system which flows from UMNO domination of the government through the Barisan National Front in Malaysia. Thus, an obvious political dimension surpassing its immediate military purposes began to crystalize around the Senoi Praaq soon after its inception and this phenomenon shrouds the RMP to this day.

Introduction and Part I: Notes

1. The Malacca Straits, the narrow passage from the South China Sea to the Indian Ocean, lies between peninsular Malaysia and Sumatra. Naturally, any power in control of this choke point or otherwise positioned to impact the flow of traffic was guaranteed great wealth. The Kra Ismuth to the north also played a significant role in early commerce as a short portage route which avoided the pirate infested waters of the Straits.

2. This term has Arabic roots and translates as "original people." The Malaysian government introduced it subsequent to independence in order to infer a collective identity upon the many different tribal groups inhabiting west Malaysia and thus to signal their rehabilitation. It is misleading, however, insomuch as the Orang Asli are not a homogenous people.

3. From a contemporaneous ethno-geographic point of view the "Malays" are said to hail from Sumatra, Java and the furthest reaches of Sulawesi. However, some scholars speculate they descended from lead elements of a migratory wave of Austroasiatic peoples who worked their way down the peninsula in the very distant past. Hence, they could have crossed over from the peninsula in route to Java, Sumatra and Sulawesi many thousands of years ago only to return by whence they came at about the start of recorded history.

4. See R. J. Wilkinson, ed., *Papers on Malay Subjects* (London: Oxford University Press, 1971), 284-287.

5. Linguistic roots of this word may be traced back to Sanskrit. Many years ago sakai literally meant ally or companion. See Marie-Andree Couillard, "Les Malais et les Sakai: Quelques Reflections sur les Rapports Sociaux dans la Peninsule Malaise," *Anthropologies et Societes* 7, no. 2 (1983): 103.

6. The Orang Asli decline followed the collapse of the Malacca Sultanate in 1699. A period of chronic civil war ensued and engulfed the Orang Asli tribes. Thereafter the term "sakai" took on a negative connotation. However, it is not in use today. Couillard, 103.

7. There is an interesting little book available about the Keris and other Malay weapons by A. H. Hill et al in which this proposition is advanced. See A. H. Hill et al, *The Keris and Other Malay Weapons* (London: the Malayan Branch of the Royal Asiatic Society, 1998).

8. The country's racial base consists of Orang Asli, Malay, Chinese, Indians, Thai and some Anglo expatriates; these ethnic groups have competing prerogatives but share little common ground from a racial, cultural, social or religious standpoint. The Chinese, Indians, and Anglo expatriates are indeed relative newcomers to the peninsula and have much less

in common with the Malay population than do the Orang Asli.

9. Jordan apparently has similar restrictions placed on her freedom of action in military affairs. Tim Goldan, "CIA Error Sent Guns to Rebels in Columbia," *International Herald Tribune*, November 7, 2000.

10. In the last analysis the Americans not only turned out to be as bad for the Egyptians to deal with as the Russians had been but they outbid them for Egyptian favors as well. In this respect, the seduction of Egypt on United States terms was an antecedent to successful conclusion of the cold war.

11. The extent of the friction between Malaysia and Singapore is reflected in the many newspaper and scholarly articles devoted to this topic. For an example see Lily Zubaidah Rahim, "Singapore-Malaysia Relations: Deep-seated Tensions and Self-fulfilling Prophecies," *Journal of Contemporary Asia* Vol. 29, Issue 1 (1999): 38.

12. The underlying problems which characterize this situation are systemic in nature and cannot be overcome by the persistent allocation of low grade human resources to Malaysia. In fact, given the tremendous geopolitical significance of this country to the overall United States position in Asia it does not behoove the United States to send anything but her best to Kuala Lumpur.

13. Word of this displeasure creeps with some consistency into the local media. For an example of negative feedback see John Teo, "U.S. Must Not Try to Enforce Democracy on the Rest of Us," *New Straits Times*, 25 April 2000, 15.

14. Ever wary of Somali type scenarios, Malaysia can attest to United States reluctance to fight opponents who are truly determined to resist. Malaysian armor was in fact present in Somalia and helped to rescue American Rangers and Delta Force operators when they were very nearly overrun in Mogadishu.

15. Jayasankaran, "Under the Gun," *Far Eastern Economic Review* 161 (March 98): 20.

16. Unlike Vat 69 and various other army commando units the more secluded Senoi Praaq can be approached only under the most unusual circumstances.

17. Chinese and Indians are also employed to this end but their rapport with the Malays is not close as is the Orang Asli-Malay relationship and, consequently, they do not factor into the Malaysian defense equation as do the Orang Asli.

18. Jakun is the traditional European term of reference for the Proto Malay tribal grouping. Its use in this context today is considered somewhat archaic and does not reflect tribal differences among the Orang Asli to the same extent as the term Proto Malay.

19. The tribal tripartite employed here, the initial classificatory scheme put forth by Paul Schebesta, groups eighteen different tribes into three broad categories of Orang Asli. This framework was subsequently perpetuated by Skeat and Blagden and eventually institutionalized by Iskandar Carey. See A. Terry Rambo, "Why are Semang? Ecology and Ethnogenesis of Aboriginal Groups in Peninsular Malaysia," in A. Terry Rambo, Kathleen Gillogy, and Karl L. Hutterer, eds., *Ethnic Diversity and the Control of Natural Resources in Southeast Asia* (Ann Arbor, Michigan: Center for South and Southeast Asian Studies, 1988), 22.

20. Speculation about Orang Asli origins ranges widely from the views of early writers like H. D. Noone and R. J. Wilkinson (drawing upon Skeat & Blagden's seminal work published in 1906—*Pagan Races of the Malay Peninsula*—the first attempt to cover systematically all Orang Asli) who ascribed to a prehistoric migratory wave explanation, to the more recent hypothesis of A. Terry Rambo postulating Orang Asli evolution in place and ascribing differences among them to their incomplete adaptation to different modes of subsistence. See Rambo, 19.

21. When Malay royalty harks back to the good old days their nostalgia is for the time when Malacca was the main commercial center in all Southeast Asia. The Malacca Sultanate is the source of a distinguishable pattern of development that marks what has become a sugeneris indigenous political system. In sum the Malacca tradition is the Malay version of Camelot. See J. M. Gullick, *Indigenous Political Systems of Western Malaya* (London: The Athlone Press, 1988), 7 and 134.

22. The British choice of Singapore island as their Southeast Asian redoubt, 125 some odd miles or so as the crow flies south of Malacca, is no coincidence either.

23. Orang Asli/Malay relations are best in Kelantan, Perak and Johore due to the residual effects of these alliances. In fact, some Orang Asli claim that some heirs to these sultanates were Orang Asli who adopted a Malay lifestyle many years ago.

24. As a general rule the Semang and Senoi in regions to the north move around (within various river valleys or traditional tribal lands) more than the various Jakun groups in the south. Southern Orang Asli live a lifestyle very similar to that of agrarian Malays.

25. Interestingly enough though the Malays and Orang Asli often recognize each other on sight despite their shared features. Such staid understanding comes from their long and intimate association with one another.

26. UMNO leads the BN, Malaysia's coalition government which retains control of the country despite severe economic problems, some internal political squabbling following the sacking of Deputy Prime Minister Anwar Ibrahim, and the strong showing of the Islamic Fundamentalist Party, Parti Islam Se-Malaysia, in the year 1999 parliamentary elections.

27. The treaty was signed on December 2nd at Haadyai in Songkhla Province, Thailand.

28. Communist combatants who fought for the MRLA were thus labeled as bandits or terrorists as opposed to revolutionaries. A favorite term of reference for MRLA fighters was "communist terrorists"—CTs for short.

29. These battalions are grouped into five brigades—Northern, Central, Southern, Sabah and Sarawak.

30. GOF now replaces the Police Field Force (PFF)/Police Gerakan Am (PGA) designations previously in use.

31. This Unit was initially designated Vat 69 after a particular brand of whisky popular among British officers. Today it is known casually as Commando 69 and designated as Battalions 1 and 8 of the General Operations Force (GOF) (the Ulu Kinta Battalions). The term Ulu means headwaters (of the Kinta river) and denotes deep jungle. The unit maintains a headquarters just outside Ipoh. Hence, it continues to dabble in jungle warfare but now operates in conjunction with an in-house Special Task Force for urban assault.

32. The Senoi Praaq is the only unit known to have been granted permission by the SAS to wear its distinctive red beret. Today though, members of the unit sport a black beret.

33. SAS operatives are also understood to have trained an elite American unit known as the Delta Force. Hence, quite a bit of their *modus operandi*, between what they gave away to Malaysia and the United States, is no longer exclusive. The SAS connection alone renders analysis of the Senoi Praaq extremely appealing to anyone with a broad interest in irregular warfare.

34. Unit insignia designed by R.O.D. Noone use the "Pra'ak" spelling.

35. Jabatan Hal Ehwal Orang Asli.

36. Persatuan Orang Asli Semenanjung Malaysia.

37. The SAS performed this function prior to the development of the Senoi Praaq as a fighting force. In point of fact, the SAS was said to have been the principal arm of the DOA. Anthony Short, *The Communist Insurrection in Malaya*: 1948-1960 (New York: Crane, Rusak and Company 1975), 450.

38. See Roy Davis Linville Jumper, *Orang Asli Now: The Orang Asli in the Malaysian Political World* (Lanham: University Press of America, 1999), 159.

39. Orang Asli political leaders generally applauded the reorganization. Some progressive elements viewed it as an opportunity to participate in Malaysia's economic development. JHEOA was reorganized to the Ministry of Rural Development in early 2001. See "Orang Asli Affairs Now Under Rural Affairs," *New Straits Times*, January 25, 2001.

40. Information from the JHEOA still reaches the Home Ministry but its source lies in individual employees not the JHEOA itself. For that matter, most Orang Asli auxiliary organizations are riddled with Special Branch informants. There are thus no secrets in an otherwise secretive community.

41. Some resistance to this idea did occur in certain quarters of the Malaysian government. However, reason prevailed and the powers that be were made to understand Orang Asli political development was inevitable. It therefore made sense to keep this process out in the open whereby it might be kept track of and better understood. Not only that, but in a soldier's vernacular ". . . it would be better to have the Orang Asli inside of the tent pissing out rather than outside of the tent pissing in . . . " and so this was a much less irksome way to go.

42. Relations between the Vietnamese and the Montagnards stand in sharp contrast to the Orang Asli/Malay relationship. Vietnamese and Montagnard truly loath one another to the exclusion of all other ethnic groups. The Vietnamese term of reference for the Montagnards is Moi or savage and hostility between the two groups detracted greatly from their cooperation during the Vietnam conflict.

PART II

The Emergency and Turn of Events Leading to Formation of the Senoi Praaq

The Senoi Praaq's deep roots grounded in many centuries of Orang Asli/Malay interaction reflect a complex array of socio-cultural, political and military forces in play on peninsular Malaysia. However, much as the history of the region itself is not hard to outline, the whys and wherefores for raising this aboriginal troop during the Malayan Emergency are fairly easy to identify, analyze and relate.

In Part II we see how the communists brought Orang Asli fighting prowess into the fray and witness the British high command's belated realization that the key to victory lay with the Orang Asli. We track the Emergency—Malaysia's long tortuous war—through three critical phases in which the MRLA struggled with British forces for control of the fringe areas following their failed battle for Malaysia's urban areas, entered into a slow decline in the deep jungle and finally retreated to some of the world's most difficult terrain along the Thai/Malaysia border in the vicinity of the Betong Salient where a hardcore MRLA remnant remained intact and poised to strike back at government security forces. Special non-conventional means had to be devised to deal with this particular problem. Hence, in Part II we take a first step toward substantiation of the premise that the Senoi Praaq warriors are an essential pillar of support in the maintenance of GOM's existing national security arrangements. Operating from the vast interior jungles and mountains they are a unique force. They guard the back door to Malaysia's cities and towns.

Chapter 6

The Struggle for the Fringe Areas

Now let us put aside matters concerning the restructure of Malaysia's security arrangements and today's high politics and go back to those dark days of the Malayan Emergency. The fighting started on the peninsula in response to MCP failure to secure a role for itself in the post-World War II government of Malaya. The communists drawn by and large from the substantial ethnic Chinese population had been Great Britain's allies in the war against Japan, and a number of British officers had enjoyed relations with Chen Peng and other top MCP leaders. Together they fought the Japanese under Malayan Peoples Anti-Japanese Army (MPAJA) auspices.[1] However, the view prevailing in the government in Britain at that time opposed any accommodation with the communists subsequent to Japan's surrender. Malaya after all was not Yugoslavia and had for that matter been a British colony in which the Chinese were a subject people.[2]

So, war became inevitable. It followed upon the heels of several years of frequent skirmishes in which the communists carried out numerous strikes, political killings and other subversive activities—both against British colonial interests and rival Chinese factions. The MCP had not been shy and essentially telegraphed its intentions, but the British paid little attention to the overall significance of these individual incidents. When the struggle finally did break out into open warfare in June 1948 it more or less caught the British by surprise. The contest progressed through two critical phases before the Senoi Praaq entered the fray.

In the initial phase of the Emergency Great Britain first struggled to wrest control of the fringe areas shy of the deep jungle from the Malayan Peoples Anti-British Army (MPABA), the largely ethnic Chinese based military arm of the MCP and successor to the MPAJA.[3] During brutal fighting in this primary phase of the conflict, rubber estates and tin mines suffered intensive attacks at the same time communist forces overran many remote government posts. By some estimates several hundred British rubber estate managers alone were murdered in the first year of the Emergency.[4]

The fighting turned nasty very early on and no European prisoners, civilian or military, were taken and local staff upon capture were themselves subject to interrogation, torture and execution.[5] Victims of torture were subjected to a variety of atrocities. Crucifixion and slow mutilation in a "Straits Chinese" adaptation of death by a thousand cuts, in which victims had the flesh slowly carved from their bodies, was not uncommon.[6] In fact the situation became so graphic the newly arrived police recruits who were rushed to Malaya and who bore the brunt of the early fighting were reported in the literature to have been instructed routinely to save their last round for themselves.[7]

Field interviews in 1999 and 2000 with veterans of the conflict confirm the reality of these instructions and establish they were hardly a matter of an officer's overly active imagination, theatrics or high drama but rather they were conditions of sheer necessity. Accounts of events on the ground published in most British sources soon after the war contrary to the face put on the evidence by others who were actually there in the thick of the fight tend to downplay the extent and extreme nature of the brutality of the Chinese insurgents. So much so their accounts—i.e. those about MRLA conduct—appear to have been voluntarily censored by the victors for the sake of good taste.

Veterans who survived the conflict considered it much more preferable to commit suicide rather than fall into communist hands alive. Rules of the Geneva convention for the treatment of prisoners were not applied and members of the government establishment, upon capture and after some preliminary unpleasantness, were lucky merely to be castrated or have their eyes gouged out and ears and noses cut off. Casualties ran high as people tended to fight to the last finger nail and close to half the above mentioned young men, British policemen sent out for duty in Malaya, were reported to have been killed in the fighting.[8]

The brief given the young police lieutenants dictated they operate solo—i.e., as the sole European—among the locals. Upon arrival in Malaya an officer was given money and guns, a map coordinate or village name, and told to go there alone to recruit a force of Malays and engage the enemy.[9] That ordinary British police officers would agree to these terms and go into the field of battle without complaint speaks very well of them indeed.[10]

THE COMMUNIST STRATEGY

From a strategic perspective the MPABA intended to disrupt the economy by crippling Malaya's economic infrastructure, to draw strength from the ensuing chaos and ultimately to take control of the country through an ever-intensifying series of military engagements. Their military plan proposed to drive British/Malay authority from the peripheral villages and towns, secure liberated zones and then mass communist forces for a push into major urban centers. The "MCP pattern of attack" was designed to engineer a sequence of events in which—once the economy collapsed, people stopped paying their taxes and government broke down—the state would ultimately yield to a communist republic.[11]

ENGAGEMENT AT KAJANG

The communist forces, however, lacked staying power due to logistical constraints, and they could not consummate victory despite their determination. Nor were their efforts to remodel the MPABA into the MRLA—a more representative and thus more effective entity—successful.[12] The decisive moment came in July 1949 during the preparatory stages of the MRLA's abortive attempt to seize Kajang, a town lying just south of Kuala Lumpur. Kajang provided communications linkage and free movement into Selangor state. In this brazen assault the MRLA suffered a trauma from which it was never to recover.

A number of men including Lau Yew, Chairman of the MCP's Military Committee and commander of the MRLA—second only to Chen Peng in the communist command structure—died during this engagement. A major supply dump also fell into government hands shortly afterwards and communist deficiencies in large unit tactics became apparent.[13] Failure to acquire this town did not bring communist losses to the saturation point but it did deny the newly constituted MRLA a center of gravity early in the conflict. In retrospect this brave group of men paid with their lives for a risky throw that, if successful, might well have brought a different outcome to the war.[14]

THE SCALE OF BATTLE

During the first phase of the Emergency MPABA/MRLA forces usually employed company and battalion sized units in an effort to overwhelm resistence but such high stake gambits to secure liberated zones lent themselves to heavy losses. MRLA casualties by June of 1952 were estimated conservatively at approximately 6,500 killed, captured and/or surrendered personnel. To continue the fight the MRLA had to replenish itself completely and drain its own resources built up so carefully during the Second World War.[15]

The general border area between Negri Sembilan, Pahang and Selangor states became particularly well known for large MRLA units on the prowl. These units remained poised for thrusts into the more populated western coast of the Peninsula and were within striking distance of Kuala Lumpur. They were well-armed and well-supplied and thus capable of mounting some serious ambushes. To deal with the problems they posed the government had to send in Gurkha companies who were eventually successful in breaking up these concentrations. Prior to Senoi Praaq formation these Nepalese troops were the best men the government had available for this type of work.

Yet, when the communists took proper precautions they were able to pick a fair-sized town and sack it before authorities could react. This occurred several times in the state of Pahang from 1948 to the early 1950s. Typically, the communists held on to these places for several days and killed anyone who was not sympathetic to their cause.[16] One such memorable occurrence took place in the town of Kuala Krau which was overrun and held for some hours on September 11,

1949. However, the MRLA could not disperse such large units quickly once an assault was over and thus it had to contend with dogged retaliatory action by the security forces.[17]

Taking and holding a lightly defended target itself was not the main problem. The MRLA lacked a modern radio communications system and this shortcoming compounded immeasurably its command and control problems. It did make remarkable use of couriers but this simple communications medium was vulnerable to security force patrolling activity, was very slow and led to ponderous decision-making.

Communist policy makers were thus obliged to contemplate changes in strategy which by the early 1950s would see them operating almost entirely from the sanctuary of Malaysia's deep jungle.[18] These adjustments in the deployment of personnel were contrary to their original plan to concentrate main force units near enemy targets and close to the many friendly Chinese settlements dotting the south and western parts of peninsular Malaya. Maintaining a significant tempo of operations in these exposed fringe areas proved to be too costly and quite a few large MRLA units of 400 or more men were badly mauled by security forces.

A CHANGE OF TACTICS

The communists had thrust Malaya into the grips of a savage war but their plan to take the country off the march had failed. A change in the communist attack strategy, in which serious incidents averaged 50 a week, became inevitable less the MCP burn itself out.[19] At this juncture the MPABA *cum* MRLA took to the deep jungle, reverting from what had briefly become a quasi-conventional force back into a guerilla-based organization. From there on, the MRLA abandoned the brash MPABA approach and adopted the same guerilla tactics the MPAJA had employed earlier in the Second World War. This strategy shift created a need for a new breed of specialized troops among the colonial forces.

In retrospect, their decision to withdraw and more fully emulate Mao's march to Yenan condemned the communist forces to failure over the long term. When they hesitated to cross the Rubicon, they forever lost the initiative and whatever chance they may have had of taking over Malaya. They formed no castle in the jungle or other enduring refuge in the hinterlands. Instead this retrograde movement changed the lines along which the war was fought and began a sequence of actions which sparked the creation of their nemesis—the Senoi Praaq.

FEAR IN THE AIR

However, British commanders had their own doubts about whether or not the communists could be suppressed and they did not rush to exploit their good fortune. They waged war to a large extent in an information vacuum of their own making and could only reckon with things they could see or experience directly. In this respect British forces suffered from two primary intelligence failures they rectified only later in the war. The post-World War II military and colonial authorities were short on knowledge about the Chinese community in Malaya. Very few British officials understood the Chinese and even fewer spoke their languages. The Criminal Investigations Department (CID) was essentially geared to fight a different battle.[20]

As the Emergency progressed British officers had to contend with an equally disquieting lack of information about the Orang Asli. This deficiency was corrected only gradually as the DOA built up its paramilitary forces. However, once the war moved through the Betong Salient area into Thailand in terminal stages of the emergency, MRLA members were reduced to their most able fighters who would not be cornered easily. Many of these individuals had become so much a part of the Orang Asli communities upon whom they depended for survival the war simply could no longer be prosecuted by the usual means. Conventional police and military methods were not effective in these areas under such circumstances.[21] The British needed on their side an indigenous force to confront the communist fighters in those remote regions inhabited by the Orang Asli.

In the meantime, throughout the early 1950s colonial authorities remained very much in a reactive mode with police and army units rushing here and there in desperate attempts to quash the rebellion. Trouble was particularly acute in the states of Perak, Johore and Pahang. Communist "traitor killing squads" had become extremely brazen and operated close to Ipoh and the prevailing opinion among local people was that the government was near collapse.[22]

Ambush killings became commonplace—even in jungle areas in the vicinity of Kuala Lumpur—further darkening the mood of British commanders. On October 6, 1951 MRLA insurgents were able in a real coup to assassinate the British High Commissioner, Sir Henry Gurney, along the Fraser Hill Road, a mere forty miles north of the capital. They managed to infiltrate the area unobserved, shot up his convoy with impunity and made a clean escape. A thirty-eight man platoon led by Siu Mah, commander of the 11th Regiment of the MRLA is credited with this feat. From documents recovered it appears as though the insurgents lay in wait along the route for several days. The event shook the British command so much so it ordered in reprisal for this action the destruction of the nearby town of Tras and the internment of some 2000 of its inhabitants.[23]

The intensity of MRLA activity did not slow much from the British perspective during 1952 even though the communists were actually running out of steam. The pessimistic view that a new general offensive was brewing gained currency. Communist pressure hung in the air like fetid tropical humidity sapping morale and detracting greatly from the true weight of ongoing counter measures.

When the MPABA began to phase out and pull back into the deep jungle the British not only underestimated their own strength but took nearly a year to understand that a strategic withdrawal was actually underway. In reality, however, reinforcements were arriving in Malaya from other parts of the British Empire and government efforts to limit support for the insurgents had begun to show dividends. The Briggs resettlement plan—devised shortly before Sir Henry's ambush to isolate the combatants from constituent support among the Chinese population—severely hamstrung the MPABA.[24] The program operating on the principle of the green desert theory aimed to interdict enemy food supply lines at their source in the Chinese squatter settlements. It was particularly effective given the MCP failure to attract outside sponsorship.

THE MRLA STANDS ALONE

Unlike in Vietnam, the Peoples Republic of China (PRC) and the Soviet Union both chose not to support actively the Malayan insurgency despite its communist bearings. And, as curious as this point may be, neither China nor Russia has given a clear explanation why no effort was made to exploit fully the opportunity for communist revolutionary warfare (CRW) in Malaya/Malaysia. Over time the lack of an outside supply of arms and food created a major deficiency on the logistic side of MRLA operations that could not be compensated for through foraging.

There is, contrary to popular belief, not that much food readily available in the jungle and large groups of men rapidly exhaust food supplies in a given area even if they are sufficiently proficient in jungle craft to know how to live off the land. The experiences of British stay behind parties operating with MPAJA units against the Japanese during the Second World War attested to the efficiency of similar but more brutal Japanese efforts at interdiction. Food supplies, the maintenance of secure food lines and access to an effective outside supply structure became a primary problem for the MPAJA and its successors, the MPABA and the MRLA.[25]

Chapter 7

The Emergency Moves to the Deep Jungle

In the interim period while British forces adjusted to this new reality MRLA units lost no time resurrecting and exploiting relationships with the Orang Asli which had been developed earlier by the MPAJA during the Japanese occupation. They quickly adopted the Orang Asli as surrogates using them once again for the same support role performed earlier by MCP sympathizers among immigrant Chinese prior to implementation of the Briggs plan and the resulting displacement of the fighting. Thus the Min Yuen,[26] the political arm of the MRLA, gave way to the creation of the Asal Clubs and many Orang Asli were drawn into the conflict.

THE ORANG ASLI GO SOUR ON THE GOVERNMENT

British operatives penned the moniker "the Asal Organization" on this new entity with its various relationships. Usage of the term "Asal" meaning original in Malay (the word has Arabic roots) signaled MCP recognition of the Orang Asli as Malaysia's earliest inhabitants. The decision to refer to the Orang Asli in this primary way had tremendous propaganda value and made it that much easier to co-opt the Orang Asli into the communist led insurgency.

Prior to this point both the British and the Malays called the Orang Asli "Sakai," a term connotative of slave or dependent. Consequently, to be conferred legitimacy had great appeal for the many Orang Asli pinned into the jungle interior where they had been largely ignored by all except a few Chinese traders following their eclipse and recession into obscurity several hundred years earlier. Under these circumstances their detached existence made them easy prey to promises from the Chinese.

As many of the Orang Asli lived in a virtual information vacuum it was not difficult for the communists to convince them the Emergency was no more than an extension of the war they had fought against the Japanese. The British would be defeated once again and driven from the peninsula a second time. Communist forces had in a *defacto* sense accepted the Japanese surrender in Malaya, as there were no British troops on hand at the time of the armistice, and they brought many of their Orang Asli supporters into the towns to celebrate the event and avail themselves of some of the loot.[27] The MPAJA dominated much of the country for several months thereafter and as a consequence gained legitimacy in Orang Asli eyes not to mention those of Malaya's more ordinary inhabitants.[28]

For all intents and purposes MRLA negotiators were already in a position of *defacto* authority and could make extensive promises to safeguard the status of the Orang Asli in a communist Malaya. Nor did they hesitate to supplement persuasion with coercion when necessary and this cinched the deal as resistance was rarely met. The Asal organization was established in short order without much difficulty and the MRLA position in the deep jungle was consolidated by 1953, well before the British could respond.

When the British did react in this second phase of the insurgency it was in conventional fashion and they wasted a great deal of time and money on large-scale search and destroy type operations before realizing these tactics were ill-suited to the task at hand. Operation Valiant—one of several large-scale operations conducted during the war—illustrates this point well. In late 1953, in this early foray into the deep jungle, 5000 troops combed the enemy infested Telom valley and Batang Padang river area for a month without killing or capturing a single communist.[29] The targeted MRLA forces, who with crucial assistance from Asal operatives, easily slipped away and lived to fight another day.

THE ASAL ORGANIZATION IN BATTLE

Thus, during Operation Valiant, and in subsequent British deployments like Operation Termite in July 1954, Orang Asli fighters took on an active role which permitted the MRLA to do battle from behind an Orang Asli screen.[30] Operation Termite per se was General Templer's first major military endeavor and designed to strike at the Temiar Orang Asli as well as the CTs in the deep jungle in the Gunong Korbu area due east from Kinta and Ipoh. This action involved a dawn air strike of three squadrons of SAS parachuting directly into the jungle. The element of surprise was achieved but the Temiar were hostile and elusive and used their blowguns.[31]

In this particular instance we see the work of the Asal organization begin to unfold. The Orang Asli acting as surrogate forces of the MRLA successfully attacked SAS troops in both the Kinta and Raia valleys east of Ipoh. They put their blowguns and poisoned darts to good use and struck down a number of troopers. The SAS can look after themselves but even they were sometimes at the mercy of the Orang Asli when in the jungle.[32]

Over the next twelve years government forces launched dozens of assaults against the MRLA. The Asal organization figured heavily into communist strategy during this period. Asal members acted as combatants, scouts and foragers for the communists and the MRLA with their help owned the jungle in a manner of speaking. The peninsula interior is quite vast in its extent stretching some 20,000 square miles up and down the main range. Such an expanded surface allowed the MRLA and their Asal auxiliaries to breakdown into so many independent units they could not be easily confronted. Within this inaccessible wilderness the MRLA was essentially impervious to conventional military operations and from the deep dark jungle it could launch with relative impunity small-scale actions reaching into the fringe areas. In this respect the operational value of tactical information supplied by Asal members was so great that it cannot be readily exaggerated.

The Asal organization, broken-down meticulously into regional groups, supported eleven MRLA regiments operating for the most part in seven western Malayan states—Selangor, Negri Sembilan, Malacca, Johore, Perak, Pahang and Kedah.[33] Each Asal organization, or club as they were sometimes called, had a Datuk Pengerusi or chairman—often referred to as Datuk Si in abbreviated fashion.[34] This individual, typically an Orang Asli hereditary leader, was used as a cutout by the communists.[35] Some say these men were mere figureheads manipulated by others from behind the scenes but such assertions do not do much to detract from Asal combat effectiveness.

THE MIN YUEN ORGANIZATION DISINTEGRATES

The Asal clubs attempted to continue the supply of men, arms, and food heretofore provided by the Min Yuen organization which had been neutralized everywhere except southern Perak and northern Johore. However, as a practical matter the relocation of the Chinese squatter population—approximately half a million people—from outlying fringe areas into new villages denied the MRLA easy access to a ready source of replenishment. Subsequent attempts at resupply under these new circumstances exposed the Min Yuen organization to Special Branch operatives who eventually penetrated and compromised the organization. To compound the MRLA problem once its troops pulled back into the deep jungle MRLA officers found it very difficult to enforce discipline on Min Yuen members and their organization left behind in the cities and squatter settlements soon collapsed. On the other hand, many of the Min Yuen's 30,000 to 40,000 Chinese collaborators were thus released for integration into the MRLA jungle based regiments providing them with their last major infusion of manpower.

Only gradually as the deep jungle fighting intensified with saturation bombing followed by raids on communist forward positions did the lower reaches of the main range hemorrhage sufficiently with Orang Asli refugees for colonial authorities to draw a clear picture of what was actually happening in the interior.[36] Through extensive interrogations of prisoners they finally understood the full extent of MRLA dependence upon indigenous elements for support and their need to do a better job breaking-up the relationship between the MRLA and the Asal organization. The only apparent solution to this problem was somehow to find a means of turning the Orang Asli against the MRLA. There was no other obvious way to root the MRLA from the deep jungle, even though it was known that such a course of action would be difficult to navigate.

BRITISH COUNTER MEASURES AND INTELLIGENCE OPERATIONS

The British had interned thousands of Orang Asli in holding camps where quite a few died from neglect before the British realized they might prove useful to the war effort.[37] Ill treatment of the Orang Asli resulted more from ignorance of Orang Asli lifestyles than from acts of retribution on the part of the British and Malay allies but the damage was done and it took some time to heal these wounds. The Orang Asli had been manipulated by the communists only to be abused by the British and they were not very receptive initially to overtures from the authorities. The solicitation of Orang Asli cooperation, the formation of the Senoi Praaq and the dispatch of indigenous elements back into the jungle as combatants against the communists was most decidedly not a straight forward proposition. It had to be worked out gradually in stages, as part of a comprehensive plan to manage the Orang Asli. This is why the decision was made at the highest levels of government to create a new intelligence service, the DOA which would ingratiate itself among the Orang Asli and ultimately woo them away from the communists. Field Marshal Sir Gerald Templer, in his capacity as High Commissioner and Director of Operations, personally decided to proceed along these lines as a part of a larger initiative to centralize the intelligence system in Malaya.

General Templer, the newly arrived commander—sent out to take over upon Sir Henry Gurney's death—was a former director of military intelligence in the British Army and held the view that victory over the communists would be won through the government's intelligence apparatus.[38] General Templer set the tone of these activities by appointing an individual with an extensive intelligence background to head up such a program—R.O.D. Noone.[39] Noone, according to a close associate, Dennis Holman, had held previously the post of Secretary to the Federal Intelligence Committee in Kuala Lumpur. A better candidate would have been difficult to dig up particularly inasmuch as he possessed a background in military intelligence as well and is also said to have worked during the Second World War in the Far East for the Special Operations Executive , headquartered in Melbourne. He was well suited to the task at hand.[40]

Work proceeded slowly at first but by January 1954 the DOA was fully operational and Orang Asli populations in the states of Perak, Kelantan, and Pahang were targeted. As a part of this process the internment camps in which many Orang Asli had been kept under surveillance were dismantled and a network of jungle outposts constructed in their stead. The intention was to take the administration to the Orang Asli instead of bringing the Orang Asli to the administration.

Operatives from the department and the SAS then used these posts as springboards into even more remote areas and slowly began the process of winning Orang Asli hearts and minds. Over a period of time they were successful in persuading many Orang Asli to relocate closer to the fort areas and to stop working with the MRLA. The provision of free medical services in particular and food supplies in general proved to be major sources of appeal to the Orang Asli.[41]

THE ORANG ASLI GO SWEET ON THE GOVERNMENT

The next step was to persuade as many Orang Asli as possible to inform on the activities of their former comrades in arms and R.O.D. Noone worked tirelessly towards this objective.[42] His efforts were exceedingly successful and by mid-1954 an exploitable situation had been created. The DOA achieved good relations with numerous Orang Asli chieftains and made many friends in the interior. From then onwards it was simply a matter of following through in order to separate the insurgents from the Orang Asli.[43]

Towards this end one hundred headmen from Pahang, Perak and Kelantan were secretly flown by helicopter to Kuala Lumpur in mid-1954 to negotiate with British officials and attend a reception held by General Templer at Karkosa—Kings House. To elicit their cooperation massive displays of British units were paraded—in a *tour de force*—before these ranking Orang Asli and prodigious sums of money offered as rewards for their cooperation.[44] For its effect this public relations *cum* diplomatic coup convinced Orang Asli leaders beyond a shadow of doubt the communists were already dead. Naturally, the assassination of prominent Asal leaders and other MRLA personalities was actively encouraged by the British who handsomely rewarded Orang Asli defectors with funds made available through the newly established DOA.[45] Thus treachery, deceit and greed were at a premium in these dealings in which many men were lured to their deaths.

The subterfuge involved in the campaign to win Orang Asli cooperation without giving the general public a realistic picture of problems prosecuting the war against the communists was such that even the DOA's true purpose was masked initially by its mundane bureaucratic affiliation with the Department of Museums and Archives. High level DOA officers like R.O.D. Noone made liberal use of their educational backgrounds in order to shield somewhat the true nature of their activities—despite their military and police affiliations—presenting themselves as anthropologists *cum* social workers.[46] R.O.D Noone's official title was Adviser on Aborigines and Director of Museums and Archives. Museum and archival duties associated with the department were not hived off until efforts to turn the Orang Asli were well underway and the Senoi Praaq deployed on operations.

A long time British Adviser to Malay royalty, Mubin Shepard, eventually took over direction of the Museum in 1958 or 1959 and became its grand panjandrum. This individual, nicknamed Fufuf, was notorious in Malaya for a public fist fight with Tunku Abdul Rahman's Australian press liaison officer—Frank Sullivan—over the affections of a handsome young waiter at the Sungei Ujong club. With independence though good times for most people like Fufuf, soon came to an end and they had to cast about for something else to do.[47]

R.O.D. Noone for his part was only too glad to be rid of these obligations to antiquated specimens and records and thus the DOA underwent its first structural change and slimmed down into a more lively and manageable entity. Hence, for all intents and purposes R.O.D. Noone had developed into a fully-fledged intelligence guru and was now poised to become a jungle warlord. Over time he became a popular demagogue among the Orang Asli, like his legendary older brother H.D. Noone who, as we shall see, disappeared in the jungle some years earlier.

THE TIDE TURNS

Albeit sanguine these subtle stratagems and ruses worked and Noone and the DOA gradually co-opted the Orang Asli populace at large and projected their influence further and further into areas infested with communist insurgents. The underlying relationship between kills and intelligence was both related and exponential. More kills led to enhanced intelligence and better intelligence provided for more kills.[48] Naturally, from a standpoint of self-interest the DOA had every reason to want its own fighting force.

DOA interference in turn led to the break up of the traditional river valley pacts negotiated heretofore amongst the various Orang Asli tribes in order to ensure that one group or another had representatives on the winning side once the direction of the fighting became more evident. In these arrangements the Orang Asli living up stream typically sided with the communists and their distant kin further down stream played along with government security forces. Fortunately for the British they were able to manifest the greater strength and once they demonstrated their superiority, Orang Asli support of the communists began to waver.[49] Nearly all the tribes soon decided to cut their losses and defected willingly to the side of government forces.

These early DOA machinations had their effect and began to drive MRLA forces northward towards Thailand and well beyond Malaya's fringe settlement areas. The trick then became how to compound these hard won initial successes with even more "active" Orang Asli participation in operations against the communists. This turn of events had begun technically sometime earlier with the formation of Special Operations Volunteer Force (SOVF) type units like the Perak Aboriginal Areas Constabulary (PAAC), the Aboriginal Auxiliary Police (AAP) and the Police Aboriginal Guard (PAG). The success of these forces, and other paramilitary type jungle squads that relied to varying degrees upon the more accessible Orang Asli, accelerated the process of winning over the Orang Asli and led eventually to the establishment of the Senoi Praaq.[50] Once this martial engineering of the Orang Asli tribes was accomplished Anglo/Malay authority over the peninsula would be virtually guaranteed—given that the genie had been let out of the bottle—if it could maintain control over both these *ad hoc* forces and the Orang Asli themselves.[51]

MACRO-MANAGING THE WAR: AND BRITISH RATIONALE FOR THE SENOI PRAAQ

British commanders were indeed chartists who generated a great deal of numeric information to feed into their decision making processes. As the war progressed they soon discovered the cost of importing personnel was not competitive with local levies of manpower. Nor were conventional forces the best means to suppress the elusive MRLA. True, by the mid 1950s the MRLA was not as thick on the ground as in the past but it continued to tie down a sizable manpower pool which at its peak some years earlier had topped out at approximately 300,000 Security Forces Personnel (British, Commonwealth and local).[52] The MRLA, exclusive of Min Yuen elements, was at its own peak in 1951 with 8,000 to 11,000 combatants. Gradually this number was whittled down and by 1954 their once ubiquitous presence was limited to three main deep jungle bases.

Thus, Communist Terrorists (CT) eliminations by security forces became harder to achieve and more costly on a per unit basis. Clearly, maintaining such elevated manpower levels involved too much overhead when fewer and cheaper specialized troops would do just as well. The resultant status quo ante meant that whenever the graphs stopped reflecting kills in exponential fashion, more Orang Asli were invariably added into the equation to thwart the MRLA. The precise origin of the Senoi Praaq may therefore be traced directly to the rejection in July 1954 by the War Office of the request by the Director of Operations (General Templer) for a second SAS squadron. This denial of General Templer's proposal led him to form in late 1956 the small experimental paramilitary force comprised of Orang Asli that has come to be the Senoi Praaq.[53] Upon the suggestion of R.O.D. Noone the formation was dubbed the Temiar Offensive Unit. Its mission: to run to the ground those hard to find MRLA squads who persisted in the rebellion.[54]

The deep jungle Orang Asli recruited into the Temiar Offensive Unit were the last untapped reservoir of manpower available to pit against the communists insomuch as a number of Malay and other integrated units had already been constituted and the creation of any more such formations was not feasible given the restrictions London placed upon General Templer's operations. Such an elastic approach which depended upon the skills of Orang Asli jungle fighters further conformed to the view that it was advisable to avoid as much direct contact as possible between British and Commonwealth troops and the local populace.[55]

Past experience had taught the British command it was best to leave people undisturbed and things as they were to the extent this was possible. The raising of an Orang Asli troop may also be viewed as an extension of the British tendency to use the police as opposed to the military to deal with the Emergency. The general idea behind their strategy was to promote synergy among the operations and activities of various security force units in order to achieve a balanced, unobtrusive and non-abusive prosecution of the war.

Having evolved a clever formula the British executed against it by the numbers in value neutral terms in spite of their concerns about the advisability of overdeveloping Orang Asli combat potential. However, the management system put in place was practical in light of the circumstances at hand; and it provided for a controllable escalation of indigenous participation in the fighting on mutually profitable terms. Already engulfed into the war the Orang Asli for their part were offered a much better deal by the government than the MRLA could hope to match. So, to their way of thinking why not profit by the opportunity? Most tribes opted to accept government overtures to cooperate once they were sure the offer was genuine.

However, some Asal leaders were extremely cagey and it took time to convince them they would not be killed or tortured as punishment for their past cooperation with the MRLA. In one instance DOA representatives negotiated with a headman from Perak for well over two years without inducing the surrender of the group he controlled. This particular surrender came about only much later when the tribal area in question became "too hot" to live in any longer.[56]

Nonetheless, despite lingering Orang Asli loyalty and indecision in some instances communist losses continued to accumulate gradually. By early 1959 intelligence assessments revealed the MRLA had sustained in excess of 12,000 casualties inclusive of those lost prior to the deep jungle fighting.[57] A fair number of these losses were incurred in a series of engagements ranging from Negri Sembilan to Perak which were designed first to pacify the middle part of the peninsula and then to sever MRLA forces in southern and eastern Johore from those operating in northern Perak. Security forces' success in clearing the central states of Negri Sembilan, southern Perak and Selangor demonstrated clearly that the MRLA, having failed to generate a critical mass in the deep jungle, was now on the defensive.[58]

Chapter 8

The Emergency in the Heart of Darkness

However, the conflict would not draw to a simple conclusion in which the MCP completely disintegrated or sued for peace. The MRLA had a tenacious core group of people left at its disposal who would fight on in a desperate effort to forestall inevitable defeat. The Emergency thus entered a third extended phase in which military operations once again shifted locus. This time the focus of the conflict moved from the deep jungle in central Malaya into the furthermost recesses of the main range along the Thai/Malaysia border in the vicinity of the Betong Salient.

Some of the world's thickest, most inaccessible, mountainous jungle terrain is to be found here as are a number of lost river valleys that only Erwin Bergot could describe and do justice. If it were not so inhospitable and dangerous to the men involved such scenery would have been truly enchanting. Some of this area remained unmapped well after the Second World War and as of the 1950s parts of upper Perak were from a practical standpoint voids in Malayan geography reaching into a true heart of darkness.[59]

While the Europeans did not know what lay in wait for them many Orang Asli believed these regions to be inhabited by cannibals and evil spirits and to some extent they were so infested. In the early 1950s, just prior to the intensification of the fighting between the MPABA and the colonial forces, Kuomintang (KMT) remnants—living off of a largely Kwang Sai Chinese population in the area—were decimated in several pitched battles which helped the communists consolidate their control over Malaya's immigrant Chinese population there and throughout the country. This battle between the KMT and the MCP was a savage, take-no-prisoners affair. Kuomintang dead were literally pursued into the grave and their livers were extracted in ritual manner then cooked in giant pots mixed with vegetables and eaten by the victors in "a typical Chinese touch."[60] When the Orang Asli unlucky enough to be present were obliged to participate in these macabre feasts as a demonstration of their loyalty to the communists their fears of cannibals were duly confirmed.[61]

These atrocious acts committed in a remote valley near the Betong Salient during the Overseas Chinese Anti-Japanese Army's (OSCAJA) last stand were recorded by British intelligence. Fighting between the MPAJA and the OSCAJA loyal to the Chinese Nationalist Government began in 1940-41 and concluded only in 1952. The combat between rival Chinese groups involved big confrontations surpassing in scale and intensity battles between the security forces and MPABA/MRLA units.[62]

The British did not support the KMT faction in these clashes insomuch as it had reverted to brigandage to sustain itself and had grown to be unreliable. British forces had in fact acquired valuable experience in several operations against the KMT in May and June of 1948 shortly before the Emergency began.[63] For this reason early on in the Emergency the MRLA was firmly established in upper Perak and Kelantan states especially around the towns of Grik and Gua Musang where it wrestled control from the weakened KMT forces. It then went on from there to consolidate its position and liquidate the OSCAJA presence in the vicinity of Benton and Ling Ong.

Thus, the Emergency was actually taking on another dimension at the very time the British presumed it was stalemated. Fewer and fewer kills were being reported and the intelligence picture was increasingly vague about the situation in this new area of operations near the Thai border. Up to this point DOA and SAS operatives had worked hand in glove to furnish the necessary human intelligence, to employ vector analysis to steer regular military units into contact with communist forces and to generate a continuous stream of kills. Mathematical analysis had taught the British to appreciate the value of small specialized units in order to overcome command and control problems during deep jungle forays.[64] The 22nd SAS regiment in fact had operated in Malaya since 1952 but there were limits to what it could accomplish in this fashion despite the refinements put in place. The SAS *modus operandi* was excellent but its fighting men were neophytes in the jungle. As outsiders they were at a distinct disadvantage in their prosecution of the war. They tended to recruit Dayak trackers *cum* combatants from Borneo rather than make use of local Orang Asli guides. To pierce this intelligence vacuum and exploit fully the existent situation it became expedient to expand the DOA beyond its intelligence function and thus to set up on an experimental basis an Orang Asli based fighting force, the Senoi Praaq. Hence, we now turn our attention to the Senoi Praaq itself and pinpoint its role in phase three of the Emergency.

Part II: Notes

1. Key among these men was F. Spenser Chapman. Chapman organized a stay-behind party to harass the Japanese and spent most of the war in MPAJA company. His story is a compelling one and well told in *The Jungle is Neutral.*

2. A parallel story to that of F. Spencer Chapman's experience in Malaya during the Second World War exists in Fitzroy Maclean's own account of life among the Yugoslav partisans allied with Tito. See Fitzroy Maclean, *Eastern Approaches* (London: Transworld Publishers, Ltd., 1975). There are of course any number of differences between their respective experiences—Chapman for instance took far greater risks—but both books are action packed and very informative.

3. There was also some trouble in major urban centers like Ipoh and Kuala Lumpur and the streets were not safe at night.

4. Herbert Andrew, *Who Won the Malayan Emergency?* (Singapore: Graham Brash, 1995), 3.

5. Edgar O'Ballance, *Malaya: The Communist Insurgent War*, 1948-1960 (London: Faber and Faber Limited, 1966), 104.

6. Harry Miller, *Jungle War in Malaya: The Campaign Against Communism* 1948-1960 (London: Arthur Barker Limited, 1972), 80.

7. There were about 400 in this first batch of reinforcements. Andrew, 23, 24 and 46.

8. Andrew, 24.

9. Andrew, 23.

10. True, there were abuses on both sides and British veterans admit to use of mild electric shock, torture by cigarette burns and summary executions. But these incidents were much more isolated and hardly a matter of routine or institutional policy. The Chinese do appear to have had a monopoly upon extreme activities of this sort as the British were anxious to encourage defections. They portrayed the Chinese soldiers as being intensely vindictive and described their actions as sheer brutality for the sake of itself. This discrepancy in levels of atrocity may perhaps be ascribed to cultural differences between peoples. For the Chinese it was a case of situational ethics wherein to their way of thinking this is just what one does to one's enemies.

11. Short, 95.

12. The MPABA became the MRLA on February 1, 1949. The idea behind the name change was to broaden the organization's appeal and acquire greater popular support. Lest

the reader be confused it is important to note that the MPAJA, MPABA and MRLA were more or less the same organization. The MCP simply made over its military wing at various times as circumstances warranted in terms of strategy, tactics and organizational structure.

13. Several dozen machine guns, a number of Sten guns, 237 rifles and 10,000 rounds of ammunition to be exact. This take represents the largest arms haul made by security forces during the Emergency and came about as a direct consequence of the fighting at Kajang. See Jackson, 28.

14. The communists also had their eye on the town of Klang, Kuala Lumpur's port. See Miller, 45,46.

15. O'Ballance, 101-102. More generous estimates of communist casualties by British intelligence operatives for the same time period placed this figure at 8,000 losses. See Richard Noone with Dennis Holman, *In Search of the Dream People*, with a forward by Field-Marshal Sir Gerald Templer (New York: William Morrow and Company, 1972), 150. Indeed, the MRLA would appear to have been in a position to absorb such extravagant losses—at least from a manpower as opposed to a political point of view. One high ranking Malaysian police source places MRLA strength, in a more recent estimate, at 14,000 men at the Emergency's onset. See Desmond Lawrence "The Aboriginal Dilemma" in R. Thambipillay ed., *God's Little Acre: Batu Gajah* (Ipoh, Perak: Perak Planters Association, 1998), 124.

16. As one young police lieutenant phrased it ". . . you would die if you stood in their way." Andrew, 4.

17. Robert Jackson, *The Malayan Emergency: The Commonwealth's Wars 1948-1960* (London: Routledge, 1991), 30-31.

18. 1955 is said by John P. Cross to be a good year in which to identify a clear change in the tempo of the Emergency. "Gone were the early days of pitched battles and casualties." See John P. Cross, *A Face Like A Chicken's Backside: An Unconventional Soldier in South East Asia, 1948-1971* (London: Greenhill Books, 1996), 34.

19. Towards the end of 1948 there had been 1,274 incidents. Jackson, 29.

20. Short, 497-498.

21. Short, 493.

22. Short, 218-220.

23. O'Ballance, 114. The British are reported to have done this sort of thing on at least one other occasion in Malaya. Robert Thompson during his service in Vietnam mentions, in a memorandum on strategic hamlets, to Major-General Le Van Kim that the village of Jenderam was razed and that about 3000 people were interned as punishment for collaboration with the communists. See Thompson to Le Van Kim, 9 Dec. 1963, PRO, BRIAM (i.e., FO) Vietnam 371/170102.

24. The Briggs plan is named after Lt. General Sir Harold Briggs who served as Director of Operations prior to General Templer's arrival and consolidation of this post into the High Commissioner's function. In the United States many know this concept as the strategic hamlets program. Like other Malayan innovations its application in Vietnam was a failure. A lengthy treatise is required to explain how this happened, but one salient reason stands out above all others. Namely, the United States and the Government of Vietnam (GVN) sought to replicate the program on a country-wide basis much too quickly and did not allow time to learn from their mistakes. On the other hand, in Malaysia the new villages implementation formula was adjusted as necessary over a period of several years. The British were so meticulous in their approach some of the men to be involved in clearing out squatter villages received special training back in Britain for this purpose prior to dispatch to Malaya. One has the impression they took this matter much more seriously than the Americans.

25. *The Jungle is Neutral* makes numerous references to food and food lines. For several examples see Chapman, 152, 240, 330.

26. Min Yuen translates as Peoples Movement. This Chinese support group based in urban centers provided the MRLA with supplies, recruits and an intelligence apparatus. Jackson, 14, 15.

27. John D. Leary, *Violence and the Dream People: The Orang Asli in the Malayan Emergency, 1948-1960* (Athens Ohio: Ohio University Press, 1995), 25.

28. Many of the major urban centers were in their clutches for several weeks as well. Leary, 11. For that matter when the Japanese did surrender formally to the British in Kuala Lumpur the British Military Authority (BMA) did not rush to establish a presence in the interior. The interim gave the MCP valuable time to exact reprisals and eliminate rivals.

29. For an excellent description of the frustrations associated with this type operation see Noone, 162, 163.

30. They were also employed on this occasion by the British and with some success. Eighty armed Orang Asli took part in the fighting and fifteen MRLA were killed. Several supply dumps and many cultivation areas were also found. See Short, 367.

31. The DOA was also involved and DOA personnel attached to Major John Salmon's SAS squadron followed in the first wave of helicopters with a small five man team including Charles H. Ley and two Temiar interpreters. Major Salmon was known as a harsh disciplinarian and famous for having his men build Burma type roads in the jungle but he got along well with DOA personnel.

32. Asal members are on record on another occasion as walking through an SAS encampment on a night foray while in search of supplies. Not many people are stealthy enough to do something like that and live to tell the tale. See Leary, 105.

33. The organization of these regiments loosely paralleled that of seven earlier MPAJA regiments each of which had a headquarters and as a rule five independent patrols. There was also somewhat more limited MRLA activity in several other states—Kelantan, Perlis and Trengganu. For a good overview of the CT order of battle see Jackson, 14.

34. The term "Pengerusi" is a Malay word and means chairman. Leary, 103.

35. The Asal clubs were particularly active among the Senoi. For a more detailed account of this organization that draws upon extensive field work and the observations of R.O.D. Noone and R.C. Corfield (Protector of Aborigines in Perak). See Leary, 101-108.

36. The intent behind these Royal Air Force (RAF) raids was to destroy cultivated areas used to supply the MRLA with food and to drive the Orang Asli away thereby denying their support to the guerillas. Short, 444. These objectives were deemed sufficiently important to prompt the British to experiment on occasion with surplus 4,000 pound HE bombs which had been used to attack German targets during the Second World War. However, 500-1,000 pound bombs were considered more effective. As for scale consider that more than 4,000 tons of bombs, 10,000 rockets and 2 million rounds of ammunition were expended during attacks on 700 targets in 1952. *Vis-à-vis* these figures and a description of various air weapons employed during the Emergency see Jackson, 72, 84. Other sources in a position to offer eye witness testimony told the author some of these munitions—i.e., the 4,000 pound bombs—blew every single leaf off any trees left standing within a radius of several hundred yards.

37. By some accounts they died by the hundreds due to the psychological shock associated with their relocation, an inability to adapt to the heat of the plains and a mediocre diet of rice and salt fish as opposed to the more varied jungle foods to which they were accustomed. Short, 443. In other accounts the Orang Asli mortality rate was judged to be considerably higher. John D. Leary cites J. Cloake's (General Templer's biographer) figure

of eight thousand while one former Director of Operations at the JHEOA fixed the number at five to seven thousand. See Leary, 44.

38. Miller, 90, 92.

39. Otherwise known as "Dick" to his friends. Contemporaries who knew the man well simply would not think of referring to him in any other fashion such was his ability to put people at ease.

40. Holman, 85, 104, 141, 149.

41. R.O.D. Noone observed that in his opinion medical aid was the key attraction. Noone, 171.

42. In his memoirs R.O.D. Noone acknowledges the criticism that he spent too much time on operations and neglected the welfare aspects of the department. Noone, 202. Dedication to duty and his obsession with the Orang Asli are also advanced as the reasons why his Australian wife divorced him. She is reported to have said that she could not compete with the Orang Asli. R.O.D. Noone married in Australia after the Second World War. There was a daughter from this union.

43. Short, 455.

44. Miller, 137.

45. Noone, 163-164. It is interesting to note that R.O.D. Noone's title was rather innocuous sounding—Adviser on Aborigines and Director of Museums and Archives—in contrast to his main work. In reality he was a colonel in British military intelligence. Holman also notes the DOA headquarters staff was accommodated in the Museum's Department Office in Kuala Lumpur. Holman, 148.

46. Sir Robert Thompson, in a point to be underscored later in the text, goes so far as to describe R.O.D. Noone posthumously as an anthropologist (the area where he had some educational background, i.e. a degree from Cambridge University) in *Peace is not at Hand*, published in 1974. In reality Sir Robert held the rank of Brigadier General and Noone that of Colonel.

47. Mubin Sheppard was an exception to this rule and he later became Tan Sri Dato Mubin Sheppard. He is credited with having done good work at Museum Negara and was a considerable scholar in his own right. His academic reputation rests upon accomplishments achieved, over a lifetime, in the study of Malay culture.

48. Short, 453.

49. Short, 449-450, 453.

50. The original SOVF units catered to the Chinese and incorporated surrendered enemy personnel (SEP). However, many Orang Asli were also available for this type of work and they enlisted in mixed units like the Home Guard as well. See Holman 205, 206.

51. It is to be noted that some mixed units recruited among the Orang Asli who were then well integrated with other Malay, Indian and European elements, as in the case of the PAAC.

52. Jackson, 18.

53. Short, 455, 472.

54. When Noone approached the SAS he initially pitched the Orang Asli to them as trackers although he no doubt had a combat role in mind for his proteges as the DOA itself had already been engaged in paramilitary activity for some time.

55. As an afterthought consider that the cumulative cost of these forces from 1948 to 1957 was 700 million pounds sterling. See, Jackson, 115.

56. Anthony Short recounts this tale and other stories in *The Communist Insurrection in Malaya, 1949-1960* which is much more than a dry history. As such, publication of his work was delayed several years for security reasons and it is by far the most comprehensive

public account of the Malayan Emergency. See Short, 454, 455.

57. Noone, 199.

58. Jackson, 53.

59. H.D. Noone was the first European known to penetrate into the mountainous area north of the Cameron Highlands. See Noone, 5-6. We also know from Chapman's exploits that even with a rudimentary map navigation is problematic in mountainous jungle "the most difficult in the world" without a point of reference or any means to gauge distance. Chapman, 63-64.

60. For additional references to cannibals and Orang Asli beliefs see Noone, 6.

61. F. Spencer Chapman himself admits to participating unwittingly in an earlier MPAJA feast along the Seroi River in which the liver and heart of slain Japanese soldiers were consumed in like fashion. From his description of the event it would appear this sort of business was indeed fairly commonplace.

62. The MCP is also reported to have had a capacity to effect the peaceful take over of non-communist gangs. Once in control the communists would then liquidate the existent leadership. This in fact happened to the Bahau (Negri Sembilan) guerillas. See Chapman, 202.

63. The OSCAJA, or at least a faction thereof, is said to have adopted a new name—the Malayan Overseas Chinese Self-Protection Corps—by this point. See Jackson, 27.

64. Jackson cites the relevant kills per contact statistics for the mid-1950s. Between 1953 and 1955 they ranged from .68 to 1.13 kills per contact depending upon the security forces unit in question. These figures signify a vast improvement in the performance of small hunter-killer platoons over larger units. See Jackson, 117.

PART III

The Senoi Praaq Takes Shape

The Senoi Praaq was set to strike the MRLA a crippling blow. Yet, it still had to prove itself in the eyes of British officials who by and large held a stereotypical view that denigrated Orang Asli combat capabilities. This shadow of doubt obliged Senoi Praaq commanders and men alike to sally forth breathing mayhem and death in order to give their unit a very auspicious start.

Senoi Praaq fighters quickly dispelled the Orang Asli reputation for docility and they wrecked genuine havoc upon die-hard communist insurgents. Ultimately, they pushed the rump of MRLA forces across the border and into Thailand. The conflict then stalemated while the MCP consolidated its grip upon three provinces in southern Thailand.

This lull set the stage for the dispatch later of a Senoi Praaq contingent to Vietnam to give a hand to the beleaguered anti-communist forces there. It also ushered in a period of bitter bureaucratic infighting between Senoi Praaq commanders and other security forces personnel. As punishment for its testy quarrels with other units, the Senoi Praaq was hived off from the DOA and assigned to the MIS.

Hence, in Part III we examine the Senoi Praaq role during later stages of the Emergency, bring its material accomplishments into the 1960s and track some bureaucratic infighting and professional quarreling which threatened to swamp the organization.

Chapter 9

Constructing the Senoi Praaq Mystique

The target date for Malayan independence ("Merdeka") from Great Britain—August 31, 1957—was fast approaching and there was a pressing need to suppress communist activities sufficiently for the British to withdraw gracefully. For this timetable to be met SAS operations had to be expedited and a comparable unit groomed to take over their role within the next few years. To this end, R.O.D. Noone badgered General Templer into acquiescence and then formed the Senoi Praaq as an arm of the DOA in late 1956.

This expansion, which followed close upon the heels of a previous surge in DOA personnel, was buttressed by an unprecedented fiscal expenditure.[1] The priority of the situation was such that eventually the DOA yearly budget would be increased to 7.6 million Malayan dollars.[2] Naturally, a great deal of its money, time and energy was given over to the development of the Senoi Praaq. For purposes of comparison consider that the overall cost of the Briggs plan which saw to the resettlement of 461,822 Chinese squatters, over a two year period into 509 new villages, was only around 41 million Straits dollars.[3]

NUCLEUS OF A NEW FIGHTING FORCE

The Senoi Praaq started out small. The initial table of organization consisted of two sections of ten men each raised from DOA operatives and Surrendered Enemy Personnel (SEP) among the Orang Asli. Two sections soon became three and the ball began to roll.

These men were mostly Asal Organization members captured in fighting along the Cameron Highlands Road by Charles H. Ley, who would later take charge of A squadron.[4] Mr. Ley is credited in particular with the surrender and recruitment of Bah Tangga, one of the more formidable Orang Asli communists. Bah Tangga, had a sizable Semai following, was no ordinary Asal member and he later became very tight with R.O.D. Noone.

To encapsulate, the capture of Bah Tangga, Mat Noor, Busu and others was by ambush involving local friendly Semai in a settlement in the Boh Tea valley in the Cameron Highlands. Charles H. Ley, O.C.P.D. Mike Abrams, eight Malay Police officers and eight armed Semai lay in wait over four days for a party of twenty armed Asal members who were literally taken by surprise and overrun and overpowered without a shot's being fired. The action was no walk in the park as Bah Tangga was armed with a Thompson sub-machine gun and the others had pump guns. In sum, they were a formidable group and at the time were unaware that later they would unwittingly form the nucleus of the Senoi Praaq.

Actually, most of the first recruits were from the Semai tribes but R.O.D. Noone dubbed the new squad the Temiar Offensive Unit because the Temiar tribes were his favorite Orang Asli people.[5] These few men were tested on an experimental basis for a time and then their number was expanded in 1957 into two full squadrons of 80 men each. The SAS under Lt. Col George Lea DSO, MBE who worked very closely with R.O.D. Noone tasked Major John Slim to train and vet these fighters as potential surrogates for the new DOA operation.[6] For that matter by this point some SAS operatives had grown weary of their own Dayak helpers whom they, according to lore, sometimes likened to taxi drivers from Kuching.[7]

JUNGLE BATTLE READINESS

When British Military officers pondered the wisdom of putting on foot an Orang Asli based paramilitary force they had to address some difficult questions measuring the fitness of Senoi Praaq commandoes for combat with the communist terrorists: Did these lithe and seemingly frail men from the jungle have the physical strength for the job at hand? Given their past alliances with the MRLA insurgents could they even be trusted? And most importantly did their benign appearance veil a ruthless will to kill?

Orang Asli volunteers who met certain physical requirements were subjected to extensive screening and they were carefully scrutinized before acceptance into the program. Their recruitment usually involved communal acceptance by existent personnel and this practice of mutual accommodation extended to prospective European and Malay members of the unit as well.

Physical qualifications were very basic. The Orang Asli recruit had to be alive, in one piece, and able to carry over hill and dale almost indefinitely a one hundred-plus pound pack. There is in this regard no comparison whatsoever between marching over open ground with this sort of load and hoofing it in the jungle.

To fully appreciate the degree to which the Orang Asli were better habituated to their environment than their new European allies consider that the average life expectancy of Europeans in the Malayan jungle without access to a variety of pharmaceutical products was only a few months to a year at best.[8] Similarly, Senoi Praaq troopers in their natural habitat outperformed the SAS operatives who were supposed to be their "patrons" by a factor of five to one thanks to their ability to capitalize upon instinct and move effortlessly through the jungle.[9]

Senoi Praaq troopers soon demonstrated their ability to bear bigger loads than either Malay or European jungle troops. An individual's backpack often approximated his own body weight, the median Senoi Praaq trooper's weight being about 100 to 115 pounds when stripped down to his undershorts. Nonetheless, the troopers were able to carry these heavy packs without compromising their mobility despite the rough, hilly, sometimes mountainous terrain. A special pack was designed later in the conflict for exclusive use by the Senoi Praaq in order to lessen the danger associated with stumbling down an incline and then being carried forward head over heels by one's own equipment. However, the ancient portage problem was not completely resolved until the specially modified backpacks manufactured along Bergen lines but without the mctal frame (to cut down on weight) were issued in the 1960s.

The improved packs enabled a Senoi Praaq trooper to carry three weeks rations (in addition to other gear) and stretch them into a month if need be. The ability to take along a fair supply of food was a critical element in jungle patrolling. The jungle harbored much less food than one might think. The need to maintain strict security left little opportunity to forage for sustenance. Air drops were the only means to secure resupply but receiving one was akin to announcing publicly to the MRLA a troop's presence in some particular area. Air drops were also dangerous affairs liable to provoke a MRLA attack to recover the fallen supplies.

The initial recruits into the Senoi Praaq were thus an unusual breed of extremely fit men already very good at what they later became famous for and not without prior experience. Violence came naturally to them and they were not shy about killing. Most came from unstable areas along the main range where the inhabitants were conditioned to jungle warfare and some had hand to hand combat experience in the MRLA.

Naturally, British officers generally viewed them with a degree of trepidation given their past dalliance with the communists and some feared the Orang Asli would be unreliable partners. These misgivings about the wisdom of arming the Senoi Praaq were compounded by mistaken, sometimes spurious notions prevalent at the time.[10] The sole European—Herbert Dean ("Pat") Noone—who lived among them and really did know them well disappeared in the jungle during the Second World War and is presumed to have been murdered (shot through the eye with a blowgun and dispatched by parang) by a Temiar.[11] Some British officers held the erroneous impression the Orang Asli were either incapable of violence or cowards predisposed towards deceitful tendencies. Neither proved the case. The Orang Asli not only were an unknown quantity, they were also terribly misunderstood and unappreciated as well.

A BENIGN MANNER MASKS THE WILL TO KILL

How could such an inaccurate stereotype of a people prevail in the minds of so many? Like most creatures in the Malayan jungle, the Orang Asli developed their instincts for survival into an art form in which success often depends upon the ability to present a misleading appearance. Actually, their innocent personalities and benign manner frequently conceal a behavioral norm in which savagery and cunning are not uncommon qualities. The Orang Asli, contrary to popular belief, do not have an aversion to killing unless it is in proximity to their settlements lest they be haunted by the ghosts of their victims. Once removed from their Saka or traditional tribal area they quickly shed their inhibitions and make very good killers. The aforementioned Semai initially handled by Charles Ley in A Squadron in particular demonstrated great ruthlessness once they took up arms for the government.

More than one European had been duped by them in the past. John P. Cross, the eccentric Gurkha commander offers belated testimony—in an account of his own experiences—to their ability to play the "rope a dope" to perfection. Cross was not fooled easily and if they could bamboozle him by feinting docility they were capable of deceiving just about anyone who does not possess intimate knowledge of their mind set.[12] So good were they at this R.O.D. Noone set up a special interrogations unit to debrief Orang Asli defectors. It was housed at DOA headquarters in Gombak and sometimes months of work were required to secure useful information out of a single suspect.[13]

However, as a consequence of such misperceptions among both British and Malay certain quarters resisted early on expansion of the DOA into the paramilitary domain but their opposition was soon overcome.[14] Orang Asli protean behavior, which British and Malay officers rightly questioned as many of them had played a double game, turned out to be an asset rather than a liability. The jungle itself may have been neutral but the men fighting inside it were not so benign. It was not unusual for Asal members to murder their MRLA sponsors for reward money and shortly thereafter join forces with the Senoi Praaq.[15] Simply put, these Senoi Praaq were mercenaries acutely aware of the changing circumstances engulfing their world. They need only be thought of in this light for their unique talents to be appreciated.

Chapter 10

Senoi Praaq Deployment

Once the decision was made to proceed the Senoi Praaq's actual startup was quick like the DOA several years earlier. Training lasting a mere three months focused on the use of weaponry and various nifty killing techniques honed by the SAS. The Orang Asli had a great fascination for small arms and they excelled in their use. They quickly learned to handle a variety of weapons with great skill. They had some prior experience with shotguns when they fought with the communists but in their new situation the Bren gun and Mark V Enfield rifle were absolute favorites.

The ambush, the mainstay of close-in assault tactics, was practiced with remarkable enthusiasm. British officers soon learned this sort of work came easily to the Senoi Praaq. Two other SAS requirements for success in jungle warfare—speed and surprise—were ingrained in the Orang Asli from birth. These particular skills allow a troop on the prowl to pick the time and place to confront the enemy and thus to stack the odds in its favor. However, for the Senoi Praaq troopers winning in combat was more a matter of how to deploy and to reshuffle their squadrons at will than the employ of these basic SAS routines.

As Great Britain's direct participation in the Emergency lessened, the Senoi Praaq warriors picked up the slack for the SAS and they soon became notorious for their accomplishments. Armed with a fresh opportunity to recycle Orang Asli talents usurped from the now compromised Asal organization, the Senoi Praaq was set to pursue its unambiguous mandate to rid Malaya's interior of hardcore MRLA insurgents. This demanding task also carried with it the painful burden of rooting out their fellow tribesmen who had remained loyal to the communists. Many of these diehards had assimilated with the Chinese through marriage. British officers described these tenacious combatants as the "tough nuts."

THE BAMBOO OPERATIONS AREA

The Senoi Praaq squadrons were gradually fed into combat alongside regular military units. At first they were not exposed to assignments too perilous. Once

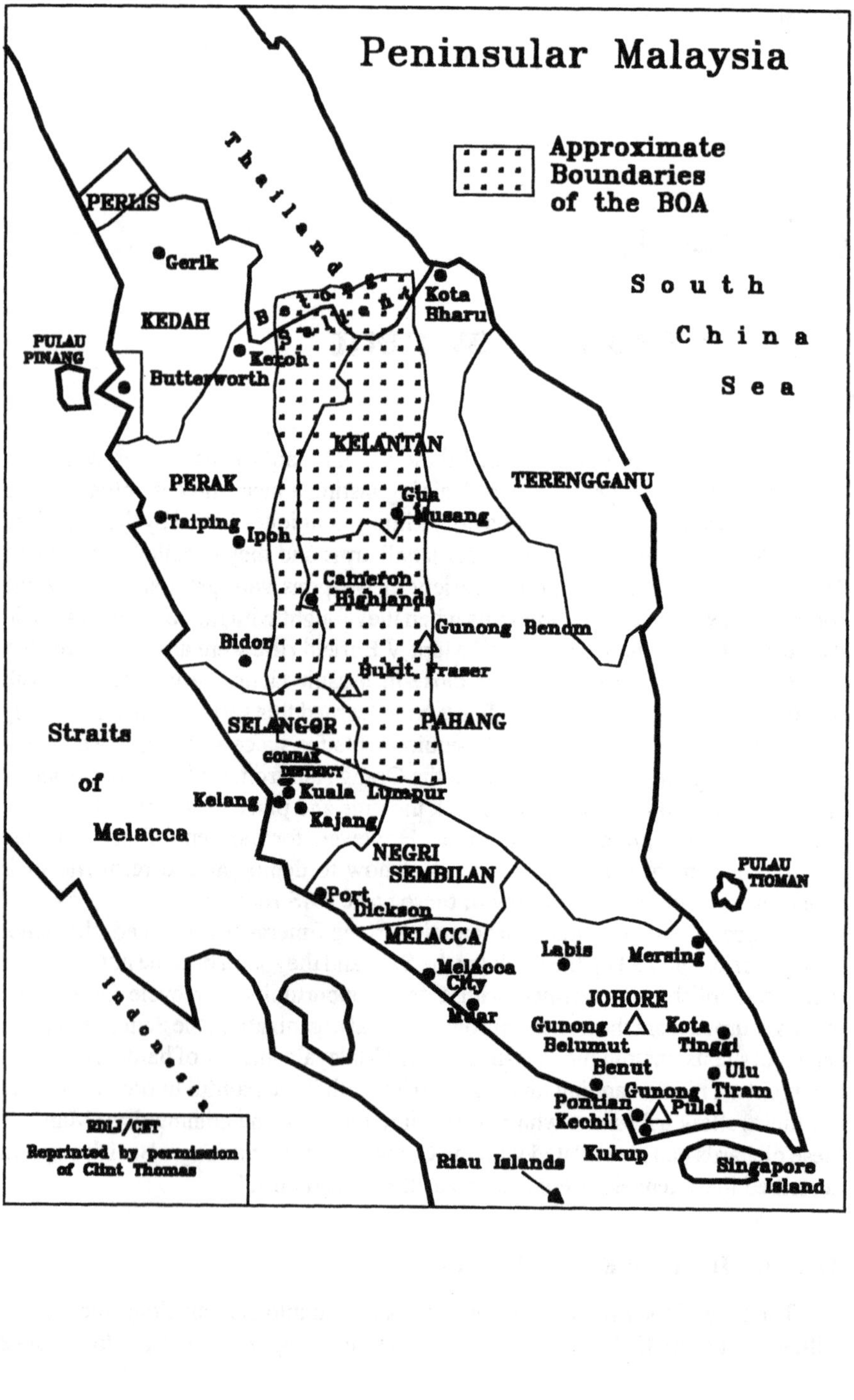

Peninsular Malaysia
Approximate Boundaries of the BOA
Thailand
Batang Sultan
PERLIS
Gerik
KEDAH
PULAU PINANG
Keroh
Butterworth
Kota Bharu
South China Sea
KELANTAN
TERENGGANU
PERAK
Gua Musang
Taiping
Ipoh
Cameron Highlands
Gunong Benom
Bidor
Bukit Fraser
Straits of Melacca
SELANGOR
PAHANG
GOMBAK DISTRICT
Kuala Lumpur
Kelang
Kajang
NEGRI SEMBILAN
PULAU TIOMAN
Port Dickson
MELACCA
Labis
Mersing
Melacca City
Indonesia
JOHORE
Muar
Gunong Belumut
Kota Tinggi
Benut
Ulu Tiram
Gunong Pulai
Pontian Kechil
Kukup
Riau Islands
Singapore Island
EDLJ/CET
Reprinted by permission of Clint Thomas

they were adjusted to their new role Senoi Praaq troopers ranged further afield in search of the more dangerous enemy diehards. The Senoi Praaq's general area of operations was—and still is—the Bamboo Operations Area (BOA) spreading along two strips of land running down either side of the main range (exclusive of the heights themselves) from southern Thailand all the way to Johore and into the coastal plains.

The MRLA tended, as we know, not to infest highland areas in excess of 4,000 feet because the MCP needed a host upon which to survive. Few Orang Asli inhabited these upland regions due to the cold weather and the absence of construction grade bamboo. However, there were some exceptions to this general rule. As a case in point security forces eventually ferreted out a base camp housing nearly 200 communist fighters atop Bujang mountain and a complex of smaller satellite camps in the surrounding area. This mountainous peak in northwest Pahang rises to 5,850 feet above sea level.[16]

Business was good but the British command eventually realized its usual force structure in which conventional troop performance could be enhanced by the help of irregular forces was counter productive when the Senoi Praaq were so employed. As it turned out the Senoi Praaq combatants were far more proficient in jungle craft than the SAS/Dayak teams they replaced, so much so they were hampered by restrictions placed upon their movements during combined operations. Tethering them to more cumbersome conventional units was a waste of specialized manpower and squandered their potential. Only late in the day were the Senoi Praaq warriors unleashed to operate alone thanks to the persistence of their commander, R.O.D. Noone, who resorted to some aggressive lobbying of both military and civilian political circles.

Once turned loose the Senoi Praaq squadrons soon managed—despite being handicapped by auxiliary duties which were thrust upon them throughout much of 1958—to secure the highest number of eliminations of any security forces unit, Commonwealth or Federal. They were employed as hunter-killer teams throughout the BOA and chalked up spectacular results in both 1959 and 1960.[17] Their kill ratio stood at 16 to 1 for killed, wounded or surrendered enemy personnel per squadron.[18] There was no longer any doubt about their capacities which took full advantage of the strategy and tactics devised by R.O.D. Noone to deal one devastating blow against the MRLA after another. R.O.D. Noone was ahead of his time in the development and adaptation of cutting edge intelligence technologies. One innovation in particular involved the production of special maps which showed rivers and contours only. They came to life under Orang Asli eyes, were remarkably good for operational purposes and useful as aids in interrogation and debriefings.

The new protocol for deploying the Senoi Praaq was straightforward. Area commanders in designated sectors simply withdrew their own conventional troops, leaving the Senoi Praaq complete freedom to operate as they chose.[19] The Senoi Praaq commander on scene would then pass a copy of his daily situation report to 2 Infantry Brigade, the entity that allotted operational areas to the various SOVFs.

Friction with other security forces (i.e. sundry Police Field Force (PFF) battalions and the Malay Regiment in particular) was kept at a minimum.

THE GOOD TIMES

Senoi Praaq personnel essentially free to range as they pleased within the BOA chose to enjoy a steady diet of CT kills. The unit from an operational standpoint operated from a high perch where it would rest for some time. Troopers thus slipped into an easy routine and proceeded to scour out the many river valleys that line northern Malaya on both sides of the main range.

Descriptions of Senoi Praaq missions offered by former British and current Malay officers differ remarkably from the portrayal offered by rank and file members. Senoi Praaq officers tended to characterize the activities of their men as rather straitlaced battles in which the Orang Asli were at their best when moving in for the kill. On the other hand, some Senoi Praaq personnel who took part in these "noble" activities professed to have been in it for the fun and the money. The more sanguine street level observations of rank and file troopers sketched a picture in which they ruthlessly hunted down MRLA remnants and killed them in a variety of bodacious ways.

Particularly memorable to the Senoi Praaq troopers were those encounters in which they used traditional weapons to pick off scores of MRLA cadres one at a time in a leisurely hunt that lasted for days. The Senoi Praaq did enjoy these skirmishes and still delight in recounting episodes about their prowess with blowguns. The crossed blowguns prominently displayed amidst their former regimental insignia remain a symbol of great pride to members of the unit and a reminder of the good times blessed by many kills.

This type killing was best accomplished when the stalker was safely concealed behind the thick new foliage found in those jungle areas that had been modified for agricultural purposes but subsequently abandoned. Such clearings typically revert back to bonafide jungle and they are extremely congested with all manner of vegetation referred to as the "beluker" stage. Visibility during peak daylight hours in virgin jungle itself is not really that bad as the canopy has choked off most of the undergrowth. However, as very little light reaches down into these areas men who spend any length of time in them quickly develop "night club" complexions.

In these situations the Senoi Praaq troopers were at liberty to employ their weapon of choice—the ubiquitous Orang Asli blowgun which is a finely crafted instrument of death—secure in the knowledge that they were free to slip away or bring fire arms into play if any particular contact became too heavy. They killed with poison shot by darts from the blowgun on a breath of air. The poison extracted from the Ipoh tree is slow acting unless a particularly strong batch happens to have been brewed and then it takes only a few minutes for the victim to die from heart paralysis.[20] This deadly substance resembles curare in that it attacks synapses throughout the nervous system. The body consequently seizes up because nervous impulses can't pass from one nerve to another. It is so toxic legend has it that anyone who goes to sleep underneath an Ipoh tree never wakes up again.[21] All in all though, death from the blowgun may be considered a merciful way to go given that there are a number of more unpleasant means by which to die in the jungle.

MRLA combatants also risked impalement upon bamboo stakes derived from lengths of living bamboo slashed through a foot or two from the base of the plant with the aid of a parang. Hapless individuals have long been found mysteriously impaled in Malaysia's deep jungle and the Orang Asli have a preoccupation with the fear of dying by such impalement. As Dennis Holman informed us in *Noone of the Ulu* "The Temiar nemesis is to fall upon one of these stumps in such a way that the bamboo enters the anus and the victim is impaled."[22]

Clearly, times had changed. Prior to the deployment of the Senoi Praaq the security forces went to great lengths to induce MRLA surrenders and to treat their prisoners humanely. Many rank and file MRLA members were dissatisfied with their lot and disillusioned with communism. Abilities and needs did not have much to do with the allocation of commodities in the jungle. The top leaders tended to arrogate the best food and all the women for themselves at the expense of everyone else. This made their men very bitter as there were not that many Chinese girls, let alone pretty ones, to begin with in the jungle. The bulk of the MRLA therefore was vulnerable to government propaganda and not a few gave up the fight out of spite for their erstwhile comrades.

Consequently, British authorities—unlike MCP leaders—did not wish to put MRLA men into situations in which they would fight to the bitter end. There were quite a few of them and if prisoners had been horribly tortured and executed in significant numbers word would have gotten out and there would have not been any more surrenders. However, the Senoi Praaq troopers were not overly concerned themselves with such fine points because those MRLA still holding out were not likely to surrender anyway. For this reason the Senoi Praaq did not take many prisoners, it was mostly kills.

Word of Senoi Praaq success against ethnic Chinese combatants quickly spread throughout the MRLA and the Senoi Praaq reputation as ruthless killers was soon established. Eventually, Senoi Praaq troopers had only to begin the patrol of an area to force the MRLA to abandon its activities and withdraw rather than face extinction.[23] MRLA combatants still gave battle to other security force units when pressed to do so but they took great pain to avoid the Senoi Praaq.

BLACK MAGIC AND OCCULTISM

For the Senoi Praaq psychoactive combat was not limited to novelty killings and the use of archaic weaponry. The magic of the occult world was employed as well. Chinese and Orang Asli alike share certain beliefs pertaining to ghosts and magic which frighten even some Malays.[24] Many Senoi Praaq fighters who believed in the power of magic in order to protect themselves relied upon incantations (*Jampi*) and wore necklace born charms (*Tanka*) purchased from *bomohs*, the traditional medicine men.[25] The operative logic here was circular in that these charms did not work unless the person wearing them really believed in their power. If the wearer died it simply meant that his faith was not strong enough. Thus, quite a few Senoi Praaq troopers were reputed to have had bullets pass through them or otherwise strike their bodies only to fall to the ground harmlessly. Needless to say, this Orang Asli tendency to have nine lives—whether it be merely reputed or actual—served to demoralize their Chinese opponents further.[26]

A more probable explanation, however, lies in random chance and the odds of a trooper being struck by a fast round. In the late 1950s and early 1960s the MRLA guerrilla fighters continued to depend upon stock piles of old "44 ammunition" (i.e., material the British airdropped to MPAJA resistance fighters no later than1944) stored in deep jungle caches. They had yet to establish themselves in southern Thailand and open up the supply channels they were later to exploit especially in Yala province. Consequently, the ammunition at their disposal was unreliable, underpowered and often it did not ignite unless treated by heat. So, prior to an operation MRLA units had to bake their bullets in sand, extinguish the fire and leave them overnight in the hot sand to drive out the moisture. Many such rounds lacked high velocity and they either were deflected by foliage or spent before striking their target.

THE INTELLIGENCE FUNCTION

There was however yet another more substantive element underlying the success of the Senoi Praaq and this was the quality of its intelligence work. The Senoi Praaq had a fairly extensive bag of tricks. Several gems germane to gathering tactical information are worthy of mention here. Senoi Praaq squadrons were designed to fragment into four troops, with each in turn throwing off four patrols. This meant that within a few days time any one squadron could get a good reading of the situation in numerous map squares while subjecting itself to minimum exposure. This maneuver allowed the Senoi Praaq troopers to stay one step ahead of the communist patrols they encountered and thus they were often in a position to disrupt MRLA movements. Another favorite technique was to locate a courier route and then to stake it out either to pick off a messenger or to engage in body-snatching operations. Their techniques for stripping bodies, removing the contents of articles of clothing or otherwise sifting around for useful information were quite elaborate and forensic quality procedures. The end objective was always the same—an intelligence bounty. Intelligence always at a premium invariably led to more kills.

Chapter 11

Trouble Among Allies

Senoi Praaq success in combat and intelligence gathering was almost too good to be true. Hence, the fact that trouble would not be long in fomenting should come as no surprise. The Senoi Praaq's fated friction with other security force units in hindsight can almost be taken as a given. The earliest tensions go back to the unit's officers chaffing at the bit to operate independently. Their restlessness presaged a cascade of events that was to see the Senoi Praaq wings clipped so to speak and its operations suspended briefly.

BUREAUCRATIC QUARRELS

Traditionally the main threat to an intelligence organization lies in the unwanted attentions of its immediate competitors inside the domestic intelligence community rather than rival foreign intelligence services of an external "enemy" per se. Intelligence organizations though ostensibly on the same side and irrespective of their greater loyalties commonly work against one another in an effort to secure a lock on the supply of coveted information to their government. The inbred conflict of interest among them surpasses healthy market place competition and creates the kind of contention in which the likes of Charles Darwin meet Nicholo Machiavelli.

This was the case in Malaya. General Templer was infatuated with the superior value of intelligence and the premium he placed upon the generation of inside information about MRLA and MCP activities. He put in place multiple intelligence gathering apparatuses which were essentially kept from fratricide by their ever pressing need to deal with the daily demands of the Emergency. Naturally a deceitful no-holds-barred struggle for survival, in which Marquis of Queensberry rules were absent, could not help but ensue as day to day pressures of the Emergency were relieved by success in the field. Once it became clear to the British/Malay commanders that the MRLA lacked the ability to surge forces in the Betong Salient area and that its bridgehead into Malaya was ill-equipped to harbor troops primed with a true offensive capacity, relations among their winning intelligence operations began to deteriorate.

In this well-managed *modus vivendi* heretofore characterized by an effective command, control, communications and intelligence system the jockeying for position quickly deteriorated into a bureaucratic bordello. Acrimonious accusations and low blows began to fly not only among members of the intelligence community but also between military and civilian government officials as well. The Special Branch, Military Intelligence, Police Field Force and DOA were at each others throats by the time R.O.D. Noone relinquished control of the Senoi Praaq in 1960. The unit's position was in peril.

Other perhaps more substantive reasons for these quarrels are lost to history but we do know a pattern of recrimination and back stabbing gradually developed. This psychological trauma eventually escalated into testy squabbling among security force officials. From Kuala Lumpur came calls in parliament demanding an explanation as to why the Senoi Praaq continued to be top heavy with European officers so long after Independence.[27] These inquiries were very shallow assaults upon the Senoi Praaq in so much as quite a few British expatriates remained in Malaya at the time and continued to exert a considerable degree of influence upon government.

R.O.D. Noone for his part was very temperamental about things when it came to matters affecting the Orang Asli and he was duly incensed by the carping and fault finding. This ingratitude for his services and for the sacrifices of his men prompted him to look further afield for new opportunities. So, he decided to take advantage of contacts he had made earlier with the Americans when they were trying to get in on the action against the MRLA and seek out greener pastures rather than linger on unappreciated in Malaya.

RAJAH JAMES BROOKE REINCARNATED

It's a shame Noone left Malaya because his success at stamping out communist insurgents in the vast jungle interior brought him legendary acclaim throughout the peninsula. He had come close to establishing himself as another Rajah James Brooke—the British adventurer granted a fiefdom by a Brunei Sultan in 1842—whose descendants held sway over Sarawak until after the end of the Second World War save for a brief hiatus during the Japanese occupation.[28] Much like Brooke's men, the Senoi Praaq operated for all intents and purposes as a private army. Noone, as its commander, had through the authority of the DOA absolute control over this important component of the military command the British bequeathed to the post-independence government of Malaya. The Senoi Praaq and the DOA together formed a powerful fusion of interest. They leveraged one another late into the Emergency and enjoyed a free run over much of the country.

The DOA/Senoi Praaq combination had become a virtual empire within the Empire. In this scenario we see a modest shoestring operation metamorphosed into a heavy hitter and major player in the British administration of Malaya. The Senoi

Praaq commandoes through the person of R.O.D. Noone had risen above themselves as most private armies are inclined to do and in the eyes of some British and Malay officials they had turned into obnoxious prima donnas.

No organization exists for long in a power vacuum, particularly not a fledgling one like the Senoi Praaq. This novel troika of force, intelligence and persona needed continued freedom of action to establish itself fully as an autonomous power center. It also required generous re-supply in order to maintain Orang Asli loyalties. Noone had to give up in the near term one or the other indispensable provision ultimately compromising his position in all three areas. Thus, in the absence of sufficient "piston" he lost a hardy gambit through forfeiture of institutional integrity in the bureaucratic wars back in Kuala Lumpur. Otherwise Noone's adventures would have exceeded not only Rajah Brooke's in Sarawak but also those of his older brother, Pat, who as a *defacto* leader among the Orang Asli had carved out a facsimile of his own deep jungle *Saka* or country during the Second World War.[29]

PAT'S SAKA

Prior to his murder Pat Noone went native, married a beautiful Temiar girl named Anjang and assimilated into the one of the key Orang Asli tribes where he became known by the honorific *Tata,* meaning grandfather or an otherwise older person. His long association with the Orang Asli began in the 1930s during a period of university field work in anthropology and continued through his tenure as Acting Director of Museums for the colonial government. After the Japanese forced the British out of Malaya he came to be a power unto himself among the Orang Asli and only yielded his position once he was weakened by malaria. There is empirical evidence in R.O.D. Noone's memoirs and in other works to the suggest Pat Noone did in fact became a Sakai Rajah of sorts.[30]

Towards the end of the war he was betrayed by an Orang Asli friend and close associate who murdered him in a dispute over his wife under the pretext that his activities risked dragging the Orang Asli into a confrontation with Japanese occupying forces. However, all Japanese attempts to run him to ground had failed as did MPAJA efforts to restrict his activities once they understood the value of Orang Asli cooperation.[31] The Orang Asli were never under any real danger at the hands of the Japanese and they hardly suffered in comparison to Malaya's coastal inhabitants.[32] Pat Noone was a free spirit who went down hard but he had his day in the Malayan sun and lives on in the minds of old school British expatriates, R.O.D. Noone's memoirs and through his own groundbreaking social anthropological research.[33]

Pat Noone's pioneering work among the Orang Asli tribes culminated in his development of a psychology for dealing with Malaya's indigenous peoples which would later make it possible for the communists to create the Asal organization. But ultimately this same behavioral stratagem also allowed R.O.D. Noone to wean them away from the communists and over to the government side through his superior application of Pat's ideas. Pat's theory held that a particular group of Orang Asli tribesmen could be controlled through careful manipulation of their collective instinct for group as opposed to individual survival. It was this well thought-out psychological technique for Orang Asli social control and cultural management, so successful in Malaya, that the British would later attempt to apply in Vietnam during the period of R.O.D. Noone's self-imposed exile and the summary suspension of the Senoi Praaq from operations in Malaya.

R.O.D. NOONE STRIKES OUT FURTHER AFIELD

R.O.D. Noone came to know some United States intelligence operatives through his negotiation of aerial—infra red—reconnaissance missions along Malaysia's Thai border.[34] So, he jumped at the chance to cream off some of his best officers and men for a special mission in Vietnam's highland jungles once Great Britain decided to support United States military initiatives in the evolving Vietnam conflict. SEATO acted as sponsor and bankroller of Noone's secretive sortie. This clandestine paramilitary action with obvious Central Intelligence Agency (CIA) ties supplemented the British government's open and direct support provided to the Vietnamese government through the British Advisory Mission (BRIAM) led by Robert Thompson.

Noone thus broke off from his Malayan engagements, leaving his second in command—Norman A. Herbolt—in charge of the unit and began to explore the possibility of an assignment for a Senoi Praaq battle group in Vietnam. And, as we shall see, indeed the Senoi Praaq force was welcome to apply its trade in furtherance of SEATO objectives in the Republic of Vietnam (RVN). There were some startling parallels between the peoples and topography of Vietnam's central highlands and the main range in Malaya. Prospects for transfer of British expertise in jungle warfare to the Americans and Vietnamese were considered excellent in light of the circumstances at the time.

HERBOLT ASSUMES COMMAND

Norman A. Herbolt commanded the Senoi Praaq from 1960 to 1962. One of Noone's men and an insider, he had previously been in charge of the Headquarters, D Squadron. Unfortunately, Noone's selection of Herbolt for promotion from within was a poor choice. He made a good hatchet man but was out of his depth as Senoi Praaq commander. As Senoi Praaq problems with encroachment from other security force units began to multiply Herbolt's efforts to protect unit prerogatives only compounded the difficulties. Norman had a tendency to fight and pick useless arguments with people he need not have bothered who otherwise would have been no threat to the Senoi Praaq. Some veteran officers attributed this ruinous habit to poor judgement and his prickly personality.[35]

Herbolt's own choice of a new second in command did not help matters much either. His choice fell upon Yusof Khan, a sinister individual whom associates described as "a nasty piece of work" capable of great damage. Those people Herbolt did not manage to mortify soon were alienated by Khan who really did excel at being ugly. Between the two of them they very nearly wrecked the organization. They were also tagged with alleged involvement in all manner of criminal activities of which black marketing, fraudulent accounting and theft of government property were but to name a few.

How did such a big cock-up manage to come about and overtake a well-run organization like the Senoi Praaq? Again, no one really knows for sure but the explanation advanced here is that all private armies tend to rise a bit above themselves. Similarly people like Herbolt and Khan somehow have a mysterious way of finding each other and under the right set of circumstances make for a volatile combination. Then in the absence of a well balanced, suave and forceful person, like R.O.D. Noone who kept a damper on things while he was around, shit just happens.

DOG EATS DOG

Some of this infighting was of a very dangerous nature. Over time several Gurkha ambush parties poaching on Senoi Praaq turf in the BOA were compromised by the Senoi Praaq who deliberately gave away their positions to the MRLA over the bush telegraph. DOA agents were sent into the Gurkha's immediate area of operations to inform the locals of their presence and the locals would then pass this information on to the communists.[36] These disclosures were indeed low blows given that in jungle warfare, once the enemy manages to get a fix on you, the hunter often becomes the hunted.

And so, by this point some of Herbolt's own allies within the DOA soured on his leadership of the Senoi Praaq. Others were actually afraid of him and what he might do. Disputes with allied security forces had simply gotten out of hand. Senoi Praaq field commanders now pushed the envelope and threatened openly to shoot wayward security force personnel who interfered with their operations.[37]

Certainly Senoi Praaq personnel were incensed over the poaching by other groups into their sphere of operations. Gurkha forays into the BOA did exceed the conventional security forces brief for dealing with enemy main force units. Once the enemy had become thinner on the ground the Gurkha's tried their own hand at guerilla warfare. Such efforts, especially those led by John P. Cross, who had in fact been under consideration for the position of Senoi Praaq commander, were a nuisance that did indeed threaten fragile agreements the DOA had negotiated with various Orang Asli chiefs. They offered flashy short term gains that might compromise a long term-bag.[38]

However, the Gurkha poaching was in of itself insufficient justification for such feuding. The delicate balance of power the DOA had heretofore worked out among the Orang Asli tribes was important to preserve but not necessarily at the expense of keeping stable relations with other branches of government. The new man in charge at the DOA, Iskandar Carey, worried about Herbolt's aggressive behavior and concerned for his own safety broke faith with the newly installed Senoi Praaq commander. He did not resist and in fact actively encouraged separation of the Senoi Praaq from the DOA in an effort to defang Herbolt. However, this restructuring in 1961 failed to resolve the Senoi Praaq problem. Transfer of the Senoi Praaq from the DOA to the Ministry of Internal Security (MIS) was a mere paper shuffle, not a real reorganization. MIS lasted only a few years. The Senoi Praaq was still independent, even more so once relegated to its own devices.

Matters only came to a head when the professional quarreling took on a personal dimension culminating in Herbolt's allegations of cowardice and homosexuality against high ranking officers at the Ministry of Defense. This accusation by the standards of the day was serious mudslinging and something or someone had to give way. And give way they did. One of Herbolt's targets, Colonel Fargus, was in a position to affect adversely the future of the Senoi Praaq and he managed to get the unit suspended from operations.[39] Tangling with Fargus proved to be a colossal blunder. Colonel Fargus was the post-independence Director of Operations, the principal staff officer in the Ministry of Defense (MOD). This position carried real clout. From the time of General Templer the Director of Operations called all of the shots at MOD.

Hence, it was not too difficult for Colonel Fargus to put up a convincing thesis outlining previous Senoi Praaq sins, some real and others invented. He aimed to destroy the unit but he failed. Such was the degree of enmity between the two men Fargus is said to have been willing to scuttle the Senoi Praaq solely for the purpose of getting rid of Herbolt. In the end, however, reason prevailed and the Senoi Praaq was not demolished. R.O.D. Noone still had a few friends in the administration and they intervened for him but at a price.[40] In a sort of compromise the unit remained on the books so to speak but the officers and men left behind in Malaya were obliged to stand down for a time and then condemned to undergo seemingly unending training exercises.[41]

So, the damage was done. The peninsula was just not big enough for Norman Herbolt and he was shipped off to Vietnam in April 1963 to join R.O.D. Noone and decompress. Neither man was out of the action for very long and both would pop back up again. In the meantime Vietnam's need for Senoi Praaq services was indeed fortuitous in light of the incessant bureaucratic infighting back in Malaya. In a happy coincidence, international macro political considerations in Southeast Asia dovetailed neatly into the micro management problem of the Senoi Praaq. Thus, while the Malayan political environment heated up, the Senoi Praaq was put "on ice," soon to be confronted with new challenges whose scope in many respects surpassed the demands of the Malayan Emergency.

Part III: Notes

1. The launching of the Senoi Praaq in 1956 followed close upon the heels of the 1953 reorganization of the DOA which saw the department's personnel rise from under twenty people to over two hundred. Many new staff were employed to run the DOA's agent network. Short, 450.

2. In his memoirs the founder of this new fighting force and DOA director—R.O.D. Noone—acknowledges the criticism that he spent too much time on operations but denies the welfare aspects of the department were neglected. Noone, 202.

3. Jackson, 20. The requisite monies did not bleed the British treasury since Malaya was a rich colony. The Malayan war effort was financed internally to a great extent unlike the Vietnam experience wherein the United States was obliged to embark upon a highly inflationary monetary policy to pay for the war. Britain had a tremendous commercial interest in Malaya that extended well beyond mere political matters. Malaya exported many valuable products in mammoth quantities to world markets. By the late 1930s Britain was already extracting yearly one quarter million tons of rubber, two and a half-million gallons of latex and eighty thousand tons of tin and other related products. Jackson, 6.

4. Wrangles with the police notwithstanding the party was soon released for long term interrogation at the DOA HQ at Gombak outside Kuala Lumpur. DOA attentions proved to be worthwhile though in this particular case. During the following months Orang Asli groups started coming in from the deep jungle and CT domination and this was helped somewhat by Bah Tangga whose voice was recorded on voice aircraft messages being broadcasted from the skies.

5. John Leary mentions in *Violence and the Dream People* that two of these sections were Semai and that only one was Temiar. See Leary, 152.

6. Training began actually in 1955 when the first ten Orang Asli men who would later form the nucleus of the Senoi Praaq became the subjects of an SAS experiment. These first recruits, all former Asal members, were attached to D Squadron of the 22nd SAS Regiment. Major John Slim, now Viscount OBE, DL, commanded and personally organized the training of the original Senoi Praaq contingent. The results of the initial trial were so promising the decision was made to push forward with the formation of an elite Orang Asli unit. Leary, 149-150.

7. The legend of the Kuching Taxi drivers gained broad currency in Malaya but SAS sources directly involved with the recruitment of Sarawak Rangers make it clear there were

only three taxis in Kuching at the time. Be this as it may, opinions of the value of the Sarawak Rangers vary among those who served in Malaya. This unit was based at Port Dickson and tasked with recruiting Ibans to act as leading scouts for British troops. Those who came from the upper Rejang longhouses are reported to have been first class. Not all were so good and some were badly handled. For his part John P. Cross cynically exposes the myth of the fearsome head hunter and suggests that they should be called instead the sneaky and cruel head hunters. The Dayak practice of taking heads tended to be centered upon easy victims, often women and children, who strayed too far from their settlements. Cross's aspersions were payback for an incident at a party in which he was nearly victimized in similar fashion by a drunken Dayak chief. See Cross, *A Face*, 161, 162.

8. Chapman, 121.

9. Noone, 201. For similar corroborative observations *vis-à-vis* cost effectiveness in particular see Leary, 150, 151.

10. These myths, fostered by longstanding Malay prejudices and reinforced by the British colonialists, have been put to rest by recent research which illustrates well that few outsiders knew the Orang Asli well enough at that point in time to assess their behavior. Leary, 152-153, 184.

11. For a gripping account of R.O.D. Noone's final efforts to unravel the mystery of his brother's death refer to Dennis Holman, *Noone of the Ulu*, with a foreword by Field-Marshal Sir Gerald Templer G.C.B., G.C.M.G., K.B.E., D.S.O. (Heinemann: London, 1958), 249-250. R.O.D. Noone later augments this initial account in a follow up work— heretofore referenced *vis-à-vis* the Senoi Praaq and other matters related to the Emergency— jointly authored some years later with Dennis Holman. Upon learning definitive news of his brother's fate R.O.D. Noone's blood ran cold. See Noone, 182-183.

12. Not only did they have him convinced the Senoi Praaq were not dangerous but Orang Asli headmen working with the communists evaded his questions with great ease. Cross, 55, 130–133.

13. R.O.D. Noone relates how he once took a group of Orang Asli to see the ocean and how this exposure helped to elicit their cooperation. A detailed description of interrogation procedure is chronicled in Noone's memoirs. See Noone, 173, 174.

14. The Malays *per se* were resistant initially to providing the Orang Asli with modern arms and they repeatedly protested this move by the British. However, lest this observation be misinterpreted it is appropriate to note also that the Malays were jealous of any favoritism the British showed loyalist Chinese. Malay leaders were very savvy politically and they were absolutely committed to the preservation of their prerogatives as the dominant people on the peninsula.

15. Leary, 106.

16. Miller, 135.

17. Noone, 201.

18. The general impression one has from review of materials published in Bahasa subsequent to Merdeka is that the Senoi Praaq are considered by the GOM to be a real killing machine. Their performance through the 1980s stands in stark contrast to that of other GOF units. See Mohamed Reduan Aslie and Mohamed Radzuan Ibrahim, *Polis Diraja Malaysia: Sejarah, Peranan Dan Cabaran* (Kuala Lumpur: Kumpulan Karangkraf Sdn., Bhd.), 299-304.

19. These areas were not comparable to the free fire zones created by American forces in Vietnam. The Senoi Praaq had access to air and artillery support but rarely used it— preferring to do their killing in a more intimate manner.

20. This brief delay was a historic problem for the Orang Asli in resisting Malay

incursions into their homelands. Malay raiders were typically armed with spears and other weapons much heavier than the blowgun. An Orang Asli could be overrun and slain by a wounded enemy who was actually awaiting the agony of death from the poisonous wound inflicted by the blowgun. The Orang Asli had no source of iron and they were hard pressed to trade or scavenge enough to make arrow heads let alone spear points. This lack of the raw material for war and their inability to cultivate enough paddy rice to sustain a large number of warriors kept them contained in the interior. Otherwise they would no doubt have raided lowland areas with great frequency and thus been in a good position to resist Malay encroachments.

21. Western medicine has no known antidote to antiaris toxicaria extract but some Orang Asli believe that eating dirt is one possible remedy to this substance.

22. Holman, 251.

23. The Chinese for their part were not in a position to cast stones at the Orang Asli over atrocities. Not only is there no shortage of accounts about the extent of their brutality but the sanity of the people involved is in question as well. The Chinese—as reported by British military personnel working with the MPAJA during the Japanese occupation—were truly barbaric. Chapman graphically describes the activities of traitor killing squads, including one group of only twenty men who assassinated and as circumstances permitted tortured to death more than 1000 people during the Second World War. The same small group of men literally killed fifty of the victims (men, women and children) with their bare hands.

24. Some Malay communists also made use of these charms and they became known as *Orang Kabal*. The term "Kabal" means tank and "Orang" stands for man. Here it is appropriate to note the Orang Asli have long sold charms and potions to *Kampung* Malays, their neighbors living near the fringe jungle areas.

25. The names for these charms and incantations vary from dialect to dialect. Their repute is such that even the more senior Senoi Praaq interviewed expressed confidence in their utility. Special munitions are the only known means to overcome the protection afforded by such charms. The Special Branch in fact had a batch of silver bullets made up for distribution to the Senoi Praaq during their attempts to deal with Rashid Maidin—the notorious Malay communist—who was reputed to have been a sorcerer. Rashid Maidin's whereabouts today have not been verified by the author: however, Maidin did survive to participate in the 1989 GOM/CPMAL surrender talks.

26. It is also interesting to note that the Senoi Praaq once employed a particular *bomoh* in Pahang to locate a secret route used by the communists. This same *bomoh* is also credited by R.O.D. Noone with helping to discover the location of a downed pilot sometime in 1960. See William Warren, *The Legendary American: The Remarkable Career and Strange Disappearance of Jim Thompson* (Boston: Houghton Mifflin Company, 1970), 195-196.

27. Field operatives suspected the informant was a disgruntled Protector of Aborigines from Pahang known as Badrillah.

28. Brooke family rule went uninterrupted, save for this brief lapse during the Japanese occupation, until 1946 when the British government forced it to relinquish control over Sarawak to the new Malayan Federation.

29. Pat was an anthropologist of some repute who became a strong supporter of the Malayan aborigines. He died tragically—shot through the eye with a blowgun—but lives forever as a popular demagogue among the Temiar. The Temiar are a Senoi subgroup. R.O.D. Noone was Pat Noone's younger brother by ten years. Pat Noone came out to Malaya some time before R.O.D. Noone and worked for the colonial government in a variety of postings. Pat encouraged his brother to join him in Malaya and later pulled some strings so

that R.O.D. Noone might find work at the Public Health Department of Perak as a collector of mosquito larva. Malaria was quite a problem in the old days and this was important work but some wags like to say that Dick or R.O.D. Noone, powerful though he was, got off from a humble start as a mosquito catcher.

30. Tom Stacy reports in *The Hostile Sun* that Pat Noone remained a feature of Temiar life well after his 1942/1943 death and into the 1950s. Stacy, a British soldier, made a trip deep into Temiar country in 1951 and at one village attended a ceremony at which he abruptly became aware that the name Noone was being chanted like a religious incantation. As R.O.D. Noone—who made good his escape from the Japanese and managed to reach Australia—did not return to Malaya until November 1952 they presumably had Pat Noone in mind. See Warren, 194. For key additional evidence supporting Pat's assumption of a traditional and exalted position—that of Mikong— among the Orang Asli see Holman, 62. The designation *Mikong* is a Thai title given by the Thais when they were still influential in Perak. Pat Noone traveled around the Main Range in style with a bodyguard of Temiar fighters. He alone had sufficient clout to resist MCP conformist pressures and took liberties among the communists other Europeans would not risk. Pat Noone's standing among the Temiar was such that the communists actually invited him to join the MCP prior to the hiatus in their relationship. Noone, 117, 123, 125.

31. Pat Noone spent a short time with MRLA forces at their Perak headquarters in mid-1942. However, he soon became disenchanted with the MCP and its interpretation of communist theory. Clearly, the Malayan communists had every intention of turning on the British at the earliest opportunity and Pat refused to be manipulated towards this end. He was also greatly offended by the barbarity of their methods. Mock trials and executions at their camps of innocent foragers who stumbled upon communist base areas were a particularly unpleasant business to watch unless you got off on that sort of thing. So, Pat left the communists a few months after his May arrival and disappeared forever not long thereafter. His body has never been recovered. See Chapman, 242, 246-247. Noone, 123.

32. The Japanese tended not to penetrate too deeply into the jungle and stayed close to the main roads. Chapman employed the term "unprecedented" to characterize his one encounter with a Japanese patrol which had ventured into the deep jungle. See Chapman, 276.

33. Most of Pat Noone's work was in fact lost or destroyed during the Japanese occupation of Malaya. However, his thesis "Report on the Settlements and Welfare of the Ple-Temiar Senoi of the Perak-Kelantan Watershed" did survive. See *Journal of the Federated Malay States Museums* 19 (December 1936): 1-85.

34. The United States did in fact perform a number of aerial high technology reconnaissance sorties for the Senoi Praaq in the Betong Salient region. However, these missions never generated much useful intelligence. For the record though, R.O.D. Noone's first substantive contacts with Americans came about during the Second World War while he was at the SOE in Melbourne. The SOE was a joint British, Australian and U.S. endeavor.

35. Herbolt's many problems were further compounded by his Chinese wife. She reportedly was a loose woman named Anne. Anne was also an inveterate gambler who had a bad habit of going around telling people she could get them a nice virgin on the cheap. Naturally, this choice of spouse would have made it difficult for Herbolt to move gracefully through the high social circles a Senoi Praaq commander might otherwise have accessed. To be fair to Anne though let it be said she and Norman both somehow managed to marry beneath themselves.

36. Cross, 101 & 124.

37. Cross, 102.

38. DOA schemes, the web of tangled alliances and dirty deals in the interior put to shame even Sir John Masterman's account of British contributions to the art of intelligence during the Second World War. For an interesting analysis of deception see J. C. Masterman, *The Double-Cross System* (New York: Avon Books, 1972). Be this as it may, John P. Cross actually complains about Orang Asli greed and having to meet their price. For details about how he operated see Cross, *A Face*, 93 and 133.

39. Here it is important to note that there was no evidence to support Herbolt's claims against Fargus and they are thought to have been little more than part of a scurrilous smear campaign.

40. Unfortunately, the powers that be were both in awe of and afraid of the Senoi Praaq. Concessions of some sort had to be made and a tidbit thrown to the wolves wherein the Senoi Praaq power position might be toned down a bit. Museums and archives – foolishly let go by R.O.D. Noone who had been happy to be rid of them – were already removed and this left only the organization's logistics tail to be sacrificed. Hence, the DOA began its arduous journey from the Home Ministry to the Ministry of National Unity and Social Development (MNUSD) stewardship.

41. Mohamed Ruslan bin Iskandar Abdullah, the commander ultimately chosen over John P. Cross to replace Norman Herbolt found his new charges in camp doing "bugger all" and made a snap decision to deploy these assets on maneuvers. This quick action prevented any real deterioration of their combat capabilities and/or unit cohesiveness.

Senoi Praaq armor on display. Reproduced by permission of the Royal Malaysia Police.

Senoi Praaq troops pass in review. Reproduced by permission of the Royal Malaysia Police.

Senoi Praaq troopers deploy. Reproduced by permission of the Royal Malaysia Police.

MRLA on the move. Reproduced by permission of the Royal Malaysia Police.

A typical MRLA ambush scene. Reproduced by permission of the Royal Malaysia Police.

Senoi Praaq troopers and the author examine a traditional Orang Asli blowgun. Reproduced by permission of the Royal Malaysia Police.

PART IV

The Senoi Praaq Deploy in South Vietnam

R.O.D. Noone, the Commander of the Senoi Praaq, was the kind of person who knew what he wanted and how to go out and get it in short order. He also had a lot of American contacts and was well aware of the deteriorating security situation to Malaya's north in South Vietnam (SVN). A stint by Senoi Praaq cadres in South Vietnam would provide the entire paramilitary unit welcome distraction during downtime back in Malaya.

In retrospect Malaysia's dispatch of Senoi Praaq cadres to South Vietnam comes as no surprise given the trouble brewing up north and the many existent similarities between the two countries. Their ethnic minority peoples—Orang Asli and Montagnard—were distantly related, their material environments were nearly identical and both governments shared common cause in the fight against communism. Senoi Praaq guerilla warfare skills were at a premium insomuch as United States special forces themselves were a fairly new entity, few in number and lacking jungle expertise.

Legitimate need for Senoi Praaq help dictated its services be sought-after as a means of shoring up the counterinsurgency program in South Vietnam. However, the opportunity to become involved in South Vietnam turns out to have been a more problematic proposition than initially anticipated. During their truncated stay in the central highlands the security situation deteriorated appreciably and unforseen events quickly diminished the effectiveness of the unit's elite comman-does.

Thus, in Part IV the Senoi Praaq take us on an adventure into the remote hinterlands of central Vietnam and to the Pleiku plateau. Through Senoi Praaq eyes we witness the failure of American efforts among the Montagnards and offer new insight into a great tragedy and lost opportunity.

Chapter 12

The Deal is Cut and the Die is Cast

By the time Great Britain lent her weight to the Government of South Vietnam (GVN), CRW had wreaked havoc on the country for many years. The Indochina states (Tonkin, Annam, and Cochin China) had already been the site of many, many battles. Vietnam's peoples and lands were well ravaged. But Minerva tarted herself back up to go into town again just like an old whore on Saturday night—thanks to United States attentions. So the Senoi Praaq were by no means late arrivals to what turned out to be a nasty encounter with something more than a ravishing young beauty.

APPEARANCES CAN MISLEAD

The seemingly mundane clashes leading to the climatic confrontation at Dien Bien Phu in the North and the equally disastrous French retreat from the central highlands of South Vietnam often involved fairly large engagements. In these actions 500 man units—Voltaire's big battalions—of the French Foreign Legion were prone to disappear without a trace. There was thus much more potential for CRW in Vietnam than Malaya where large unit engagements gave way to guerilla warfare in the more traditional sense of the term.

Yet, this sorry state of affairs was somehow obscured by bureaucratic obfuscation and it was conveniently overlooked by those anxious to get in on the action. With the conclusion of the French phase in the fighting, and partition of Vietnam into two separate countries, South Vietnam (essentially Annam and Cochin China) emerged as a viable political entity from the standpoint of United States operatives. However, fallout from their failure in 1954 to factor into their strategies the true strength of the Viet Minh (especially its local roots, number and capabilities) in the southwestern delta region undermined their efforts later to build up the Montagnard people in the central highlands as a protective shield against Democratic Republic of Vietnam (DRVN) infiltration along its essential supply line, the famed Ho Chi Minh Trail.

The Geneva Accords obligated the DRVN (essentially Tonkin) to repatriate all its forces from south of the 17th parallel to the north. However, at the southwestern staging center near Soc Trang, under the not so watchful eye of the International Control Commission team delegated to supervise the withdrawal, the wily Viet Minh substituted new recruits for hardened combatants who stayed behind to fill the power vacuum created by the departure of French forces. Many of these veteran communist fighters and political agents (Can Bo) were natives of the area; others had assimilated to the local milieu from years of struggle first against the Japanese and later the French. They held sway over most of the area except for some towns where French forces kept them at bay with the help of the notorious warlord (the French dubbed him"Ba Cut") who dominated the Hoa Hao Buddhist sect.

For their part the French regulars and Foreign Legion withdrew from the delta area as stipulated in the Accords. This left "Ba Cut" to stand alone. He was a real "sweetheart" and one wonders why no deal was made with the Hoa Hao. American prejudice towards the French is as much to blame as lubricous self-interest on this score. In any event, the RVN army, urged on by some of the first United States military advisors to arrive in country together with CIA personnel already on the scene, baited Ba Cut to mount an assault on Saigon where it easily defeated his main forward force. This ill-fated venture in the April 1955 battle for Saigon brought to a dramatic end French mastery of South Vietnam's fledgling armed forces and thereby closed out what was left of the French military position; but it gave Viet Minh battalions 307, 310, and 313 a relatively free hand back in the southwest to consort with rebellious Hoa Hao elements.

The RVN belatedly attempted to address the festering insurgency in the southwest delta region massing population groups into "strategic hamlets" for their mutual protection. However, Viet Minh agents (Can Bo) quickly disrupted the hastily constructed hamlets. When the already thinly stretched Army of the Republic of Vietnam (ARVN) forces in the IV Corps region failed to stem the increased tempo of the insurgency, the RVN had to divert to the southwest some of the very forces whose presence had become critical to the security of the central highlands.[1] This redeployment to the delta region came about just in time to complicate the British sponsored Senoi Praaq mission with the Montagnards.

Be this as it may, once these remedial measures had been put into place United States operatives and their RVN allies closed their eyes to the residual strength of the indigenous communist cadres who lived off the population in the southwestern delta region. As late as 1959 they continued to depreciate the Viet Minh's eerie presence, fixing the number of their fighters in the whole of South Vietnam as only about 3,000.[2] These numbers were thought to be manageable and comparable to those confronted by security forces in Malaya.[3]

Such minimalist projections of communist power persisted for several years and were adhered to by the British as well despite a conservative overall outlook on their part.[4] At the command level their own mental picture of the opposition did not change much through January 1963. Foreign Office correspondence makes it clear the British estimated enemy depth at only about 25,000 men.[5] The cumulative strength of British and American illusions thus made it that much easier for their governments to slip into a "quicksand war."

THE GROUNDWORK

In Fall 1961 Great Britain set up a military advisory mission in Saigon and put into place a mechanism for dispatch ever so quietly of a contingent of Senoi Praaq troopers into the Montagnard areas of central Vietnam. The initial official arrangements were made between Saigon and Kuala Lumpur subsequent to a formal personnel request in early 1960 by President Ngo Dinh Diem of South Vietnam to Tunku Abdul Rahman, the Malayan Prime Minister. President Diem asked that a team of experts be sent to review the security situation and advise the GVN about various measures which might arrest South Vietnam's deterioration. Sir Robert G. K. Thompson, the permanent Secretary for Defense in Malaya at the time, led this team and his report so impressed President Diem that he later asked the British and Malayan governments if he could have him back on a more permanent basis.[6]

Additional high level encounters took place throughout the remainder of 1960 between British and South Vietnamese officials and paved the way for dispatch of a more permanent mission. Field Marshall Sir Gerald Templer (British High Commissioner in Malaya, 1952-1954) made a visit to Vietnam in October and met personally with President Diem. This summit was a follow-up to extensive working discussions held—subsequent to Robert Thompson's initial visit —between General Richard Hull, Commander-in-Chief of Far East Land Forces of the British Army (FARELF) and Brigadier General Nguyen Khanh of the Armed Forces of the Republic of Vietnam.[7] Views were further exchanged and a strategy began to crystalize in which certain techniques from Malaya might be applied selectively in South Vietnam given that there was some similarity between the two environments.

THE BRITISH ADVISORY MISSION (BRIAM) JELLS

R.O.D. Noon's preliminary survey of Vietnam's indigenous peoples during this period conformed to the view the Senoi Praaq experience in Malaya could be replicated in Vietnam. He also identified several areas in the central highlands appropriate for the Senoi Praaq to set up a Mountain Scout type program (the telltale sign of the SAS signature is often identified through usage of term "scout" as an activities designation).[8] The 22nd SAS Regiment in Malaya was in fact known initially as the Malayan Scouts.

Both Vietnam and Malaya had substantial indigenous tribal populations. The French had some success in their own dealings with the Montagnards in the same time period the British cultivated Malaya's Orang Asli. Their Maquis type programs paralleled the odyssey experienced by many young police lieutenants from Britain in Malaya and at one time they had many highland peoples under arms.[9] So, there was ample reason to believe in the success of some sort of further effort to tap into this tribal resource. In this respect, R.O.D. Noone's recommendations were crucial in securing the approval of British officials for aid assistance to the Diem government although they had a number of other issues on the table. On balance, the similarities between the two highland environments inhabited by tribal minorities were felt to outweigh any problems that might come from their differences. Amusing anecdotal evidence of Noone's initial approach to the Montagnard tribes may be found in a United States Department of State report dating to 1960 in which hereditary Montagnard leaders were curiously found to be in possession of very large quantities of Tiger Beer.[10]

However, on a more somber note the initial impressions of the British leadership were guardedly optimistic and some saw only a limited window of opportunity for the transfer of their Malayan experience to Vietnam. South Vietnam unlike the elongated Malayan peninsula was not easily isolated from outside influences. The human environment was porous and in contrast with Malaya it had no large and virulently anti-communist ethnic group. The Malays represented nearly half Malaya's population at that time and, as a general but not inviolable rule, their adherence to Islam taught them to be anti-communist to the core. Nor were ethnic rivalries in Vietnam as easy to draw upon and the religious opposition to communism was fragmented. There were Buddhists, Catholics and various heretical religious sects involved—the Cao Dai and Hoa Hao were the largest—and they were perpetually at odds with one another.[11]

Robert Thompson—a little more optimistic than some of his colleagues—nonetheless opted to take R.O.D. Noone's lead and make the transition from Malayan to Vietnam specialist. He returned to Saigon in September 1961 as head of a small contingent known as the British Advisory Mission (BRIAM). BRIAM, in addition to Robert Thompson, consisted of four police officers, a clerk and two typists. Its budget ran at about 100,000 pounds sterling a year inclusive of salaries. A token sum given a United States expenditure on the order of about a million dollars a day during the same period. BRIAM thus had a diminutive profile, focused its official advisory effort on the construction of strategic villages in the ten provinces surrounding metropolitan Saigon, and served as an alter ego of sorts to the low-key British diplomatic mission in Vietnam. Great Britain's Foreign Office went to great lengths to project a detached public image of her Majesty's Government towards antagonists in the conflict. As a consequence R.O.D. Noone and his outfit were not carried formally on the BRIAM roster even though a number of advisors and some regular Commonwealth troops did ultimately serve openly in the war.[12]

SEATO'S SECRET

Senoi Praaq activities in Vietnam were subsidized by and executed under the auspices of SEATO where BRIAM could become a legitimate affiliate. Although the Senoi Praaq had SEATO sanction, technically the Malayan government never, in point of fact, became a SEATO member. Nor several years later upon formation of the Federation of Malaysia did the "Malaysian" government toe this line either. Malay distrust of excessive American influence over the organization precluded such an alignment by the GOM. GOM leadership was also wary of provoking the Peoples Republic of China (PRC) which as we know had not previously supported directly either the MCP or MRLA. Thus, there was no reason now that the Emergency was finally under control to provoke a change in PRC foreign policy.

Not only were Senoi Praaq activities kept secret to preserve the mission's integrity they were also considered to be, from a politico diplomatic standpoint, very delicate. SEATO trappings for this particular venture were really a smokescreen for CIA involvement.[13] The CIA, now widely acknowledged to have been intimately involved in the Montagnard initiative during its incipient phase, was the prime mover behind what later came to be known as the Civilian Irregular Defense Group (CIDG) scheme of which there were several related components inclusive of the Mountain Scout Program, Combat Intelligence Teams, Border Surveillance Teams, etc.[14]

Hence, the Senoi Praaq did not go into Vietnam as a regular paramilitary force. Members of the unit slated to make the trip had to decommission prior to departure and travel incognito as ordinary civilians. This obfuscation of their affiliation was not only necessary to distance the GOM from the operation but it was also in keeping with the type of cloak-and-dagger work R.O.D. Noone enjoyed. The operation, conducted off the books and shrouded in secrecy, appealed to a certain type of individual of whom there were many in the Senoi Praaq.

As a unit the Senoi Praaq already had a reputation for risky behavior but acceptance of this assignment demonstrated a penchant for living dangerously few men possessed. The penalty for failure was torture and certain death at the hands of an implacable foe. The Chinese reputation for brutality in Malaya was extreme but the real sadism of Asia was to be found in Vietnam. We know from the accounts of men like Bernard Fall, Lucien Bodard and Erwin Bergot just what it meant to be taken alive by the enemy. In such an event their Senoi Praaq connection would, upon discovery, be denied by GOM and their sacrifice ignored. Yet, despite the obvious perils of such an undertaking none of Noone's selectees took French leave. They loved the life and war was literally in their blood. The underlying implication for these tough terms of engagement and the various deceptions involved infer true mercenary status on the Senoi Praaq who served in Vietnam. Once in Vietnam they operated as individual contract agents without even a thin veneer of legitimacy heretofore afforded to them by their paramilitary designation and DOA affiliation. They were in other words completely expendable.

BRIAM

Thus, BRIAM pursued its mission along two parallel tracks—one conventional, the other covert. In sum BRIAM discreetly funneled Commonwealth men and material into Vietnam; performed an analytical and advisory function; scheduled courses for GVN personnel at the Jungle Warfare School (JWS) in Malaya; and loosely coordinated Vietnamese covert action programs for the Government of South Vietnam.[15] Robert Thompson was a natural for this post given his recent high-end managerial experiences with irregular warfare in Malaya —not to mention his earlier service as a young man during World War II in which he was notorious for leading from the front with Tommy gun in hand. Britain's new proconsul to Vietnam had been an RAF liaison officer with Wingate's Chindit operations in Burma against the Japanese, and he was no stranger to guerilla warfare in the field.[16]

However, on this particular occasion he took R.O.D. Noone *et al* along to do the dirty work of the mission and to be liaison with Vietnam's indigenous population. Thus fate thereby conspired through Noone's draft for service in Vietnam to keep the Senoi Praaq from being put out of business altogether. Indeed this excursion was a welcome opportunity that provided for a period of rehabilitation and breathed new life into the Senoi Praaq troopers. The unit was to gain international exposure through this assignment and acquire a more worldly grasp of the counterinsurgency game.

So, while Thompson, who would later be knighted for his service, hobnobbed with the Saigon elite and served as a suave front man, R.O D. Noone and the off-the-record Senoi Praaq contingent set up shop in the hills and jungles of Pleiku province.[17] The Senoi Praaq, initially headquartered at a Vietnamese army base in the vicinity of Pleiku City proper, wasted no time following up R.O.D. Noone's initial overtures to the local Montagnard leadership.

Chapter 13

The Senoi Praaq Hardcore Makes Its Way to South Vietnam

R.O.D. Noone recruited a small group of volunteers from among his Senoi Praaq squadrons for immediate dispatch as an advance party to South Vietnam in March 1963. Noone together with his squadron commanders Charles H. Ley, Zainal Abideen Alsagoff, and Mat Noor led a group of some twenty odd battle hardened British, Malay and Orang Asli which included three radio operators and again as many medics not to mention a dozen or so gunmen *par excellence* like Bah Tangga. The others along for the ride were Sunnar, Raman, Neville, Ismail, Ilias, Gaizun, Bah Ragum (medic), Tony Brown (doctor), Malcolm (wireless operator) and Monte Rodolfo, a British liaison officer. Thus, Noone had with him the specialists necessary to build up three new squadrons out of Montagnard levies. Technicians aside, they were also his best killers and most came from the ruthless Semai based A squadron.

THE MONTAGNARDS

The term Montagnard stems from the French word for mountaineer and is generic as is the Orang Asli designation. Some thirty-odd tribes inhabit the Annamite mountains which form the spine of Vietnam and are referred to collectively by this term: The Bahnar, Jarai, Kho and Rhade are only a few. In a macro-cultural sense some of the tribes resemble the Malaysian Orang Asli. In point-of-fact the Bahnar, Jarai and Rhade tribes, an Austronesian *cum* Malayo-Polynesian type people, remained behind during prehistoric migrations southward which eventually took some of their number to Malaya.

Unlike the composite Orang Asli tribes and those Montagnards further north in the DRVN the Montagnard groups in the central highlands are fairly homogenous. However, Montagnard and Orang Asli (from Malaya's main range) alike were hardy semi-nomadic peoples who—once provided with good weapons and proper direction—made excellent fighters. Robert Thompson stressed these connections in one of his books about Vietnam in which he euphemistically referred to Richard Noone as a well-known Southeast Asian anthropologist.[18] Noone for his part found Vietnam's Jarai in particular and Malaysia's Senoi to be closely akin in culture and language and noted they could still speak to each other.[19]

More specifically they also practiced an identical form of slash and burn agriculture geared towards hill rice not to mention various tubers and in this way pursued their respective modes of physical existence in much the same manner. Semlai[20] drinking rituals in Malaya are exemplary of these parallels and hardly vary from the habits of the Montagnards. Both consume rice wine from earthen jars through bamboo straws while seated upon mats in the center of a long house. Custom dictates that whenever a man becomes so drunk he begins to stagger around, one of the women seated around the walls of the structure will push him back into the center so that he may resume drinking until rendered unconscious. Alternatively one may opt to abandon the bacchanal and give up the drinking straw to an older woman of the tribe. Failure to understand that substitutes were allowed and not considered a loss of face resulted in many a needless Green Beret hangover.

No doubt such great similarity between the two peoples helped to prompt BRIAM's decision to designate the Pleiku plateau—inclusive of Pleiku and Khontum provinces—as the Senoi Praaq area of operations. The locus of Senoi Praaq operations, an area mainly populated by Bahnar tribes (75,000) and the Jerai (150,000), straddled the central highlands of SVN in line with Qui Nhong and the Pleiku plateau.[21] This Montagnard ancestral homeland amidst the central highlands was Vietnam's vulnerable backdoor just as Malaysia's great main range provided covert ingress into its coastal plains. So, a comparable geographical dimension was also in place and this in turn amplified the attendant demographic opportunities.

Over the next six months the Senoi Praaq were active north of the Cheo Reo district particularly along small rivulets leading into the valleys of the Ayun and Ba Rivers which were a near perfect match for the terrain in northern Malaya. The Ba (Dak Pa) River pierces the Truong Son mountains to the east of Pleiku and empties several hundred kilometers further south into the South China Sea at Tuy Hoa. Its valley and that of the Ayun River, actually a major tributary of the Ba river, serves as a fluvial link connecting the central highlands with the heavily populated coastal plains. This route led towards Nah Trang, the sleepy coastal town about a 100 kilometers south of the river's mouth that became the staging area for U.S. special forces operations in Vietnam.

The aggregate Montagnard population ranged from eight hundred thousand to one million people spread throughout the central highlands. Pleiku province alone —with an area of approximately10,000 square kilometers—had an estimated 187,000 people. Thus, although no one knew exactly how many hill people there were in Vietnam the consensus view was that their number was fairly large. A favorable demographic picture from a manpower perspective could not help but add further appeal to BRIAM planners. Thus, as the Jarai proper accounted for about ten percent of the total Montagnard population and nearly 75 percent of the Pleiku plateau population itself their homeland became a target of R.O.D. Noone's attention.

Consequently, Montagnard strength in numbers, racial similarity and overall compatibility—both cultural and geographical—made for a short Senoi Praaq learning curve. Quick Senoi Praaq acclimatization to local conditions would virtually guarantee—to the British way of thinking—that a fair number of Vietnam's indigenous peoples could be co-opted. And so, the Pleiku plateau became a fulcrum in the overall scheme of BRIAM planning. So much so that Robert Thompson himself often said that Vietnam could be defended against the DRVN forces as long as the Montagnards didn't go sour.

ENTER THE SENOI PRAAQ

The Senoi Praaq, in a slight departure from their former role in Malaya, were intended to be neither combatants nor assassins in the strictest sense of either term but rather they came to teach GVN forces the ways of the jungle and to make friends with the Montagnards. They had a two-fold mission: train GVN troops to function effectively in the highlands region and establish an intelligence gathering network among the Montagnards. The general idea was to marry competent GVN forces with sympathetic Montagnard tribesmen and use them to seek out the Viet Cong and interdict their infiltration into South Vietnam. In a nutshell the plan was to take over their designated operations area and cut SVN in half with Montagnard Senoi Praaq-trained forces. The men on the ground in early 1963 thought this to be possible and perhaps the dream could have been realized just as the Malay peninsula had been dichotomized during the Emergency.

In a first stab at the job R.O.D. Noone's team moved away from the shelter of the big guns around Pleiku city and together with a Vietnamese special forces squad from the Luc Lang Duc Biet (LLDB) and a Montagnard military team of 35 Jarai from Cheo Reo district (expanded to 180 men in the following weeks) set up their forward HQs in Kan Nak and An Khe and began to operate.[22]

Unfortunately there then followed a period of some months peppered by frustration and delays during April and May when it was decided that the old French fort sites (which incidentally had been overrun several times in past battles with the Viet Minh) were unsuitable. The team(s) then moved on to Plateau GI (Guard Indochine), another old French fort site set in the highlands 60 miles north east of Khontum at an altitude of about 3,800 feet. This camp site was situated near the HQ of 41 VN Infantry Regiment and a mile from the local district Civil Guard HQ and in itself comprised another potential stronghold.

Thus, the Senoi Praaq ensemble was not designed to serve as a static mechanism by which substantial irregular forces might ultimately be levied. Robert Thompson was indeed preoccupied with the break up of communist cells and the penetration of their networks of agents and he saw value in the levy of indigenous forces. However, these more sanguine aspects of the BRIAM counterinsurgency program drawn up for South Vietnam were not applied initially among the Montagnards. Such a course was not considered suitable at the time for political reasons. President Diem and his brother, Ngo Dinh Nhu, did not trust the Montagnards and they did not want to see too many of them organized, armed and given a stand alone combat capability.

Nor did generic levy of Montagnard forces necessarily conform to British ideas about the best way to secure the central highlands. Several considerations prompted the British to be selective in their planning of the Senoi Praaq assignment which was also geared towards interdiction of the infiltration route southwards (the Ho Chi Minh trail) that abutted the highlands plateau. Most notably, the central highlands despite some heavy engagements between the French and the Viet Minh was nevertheless a fairly stable region of South Vietnam. Hence, there was no strategic need to push Montagnard recruitment beyond tolerable political limits.

Peaceful areas in which the inhabitants were relatively detached from the trauma of war did not make for fertile recruiting grounds where ideological discontent could be sewn. Similarly, at the time of Vietnam's partition in 1954 the Viet Cong had scant underground organization in the area upon which to subsist and therefore not much guerilla activity could take place in the vicinity of Pleiku. The place had not, for a variety of circumstances, yet gone bad and there was no reason to overreact.

However, this state of affairs would gradually change as more and more refugees from North Vietnam were funneled into the plateau region. Hence, the Montagnards would not be recruited in strength until the United States buildup really began to accelerate a few years later. Similarly, the DRV government would also have to seize upon the significance of a border control strategy at the juncture of Vietnam, Cambodia and Laos before things really began to heat up. One thing would have to lead to another before there was a real "slugfest" over the highlands.

DEALING WITH THE BAD BLOOD BETWEEN THE MONTAGNARDS AND ALL VIETNAMESE

Once established on the ground the Senoi Praaq team had to contend almost immediately with problems created by the Montagnard's strong aversion to all Vietnamese. Senoi Praaq relations with their Vietnamese special forces counterparts also deteriorated badly because LLDB personnel continually abused Montagnard troops and Orang Asli Senoi Praaq troopers alike in various ways.[23] R.O.D. Noone personally had to warn the Vietnamese through the person of General Khanh (Cmdr. II Corps) that they should desist from this derisive behavior.[24]

Thus, the Montagnards by tradition and rightly so wanted little to do with the Vietnamese. Whether they came from the north or south, were communist or capitalist, did not matter. Ancient ethnic rivalries forbade close interaction between Vietnamese and Montagnard. The traditional Vietnamese term of reference for the Montagnard was *Moi* or savage. Unlike in Malaya, where the Malays subjected the Orang Asli to a more pedestrian form of patronization, the Vietnamese had and presumably continue to have no use for the Montagnards. At the same time the waves of Vietnamese intruders suppressed and ostracized the Montagnards alternatively ad infinitum as circumstances permitted, the Malays for their part pursued a traditional policy of matrimonial alliances with the hill tribes in order to co-opt them.

Vietnamese discrimination against the Montagnards had a collective quality in which northerners and southerners alike were equally contemptuous of the mountain tribes. Just as the GVN had trouble managing the Montagnards in its territory so too did the DRVN initially fail to make many converts to communism. In the early 1960s there was not an overbearing Viet Cong presence evident either in Pleiku or Khontum province. The Senoi Praaq were somewhat ahead of the curve in this respect and at the onset they had a fairly stable platform from which to operate. In sum, there was indeed a strong possibility of success for their endeavors if cat fighting between their charges could be kept in check.

THE VIET CONG AND THE FAILURE OF RESETTLEMENT TO CUT THE GORDIAN KNOT

Viet Cong activities were an immediate problem, however, in the hastily established settlements created through the strategic hamlet program sponsored by the Diem government. The intent of this program—a degraded version of the plan implemented earlier in Malaya as a part of the Briggs plan, attempted to establish a buffer zone of loyal Vietnamese in the highlands. Monsignor Ngo Dinh Thuc, President Diem's elder brother, organized an exodus of some 800,000 Catholics from the DRV upon French withdrawal from Indochina. Many of these refugees eventually were relocated in the new security settlements. Unfortunately, a fifth column infiltrated the refugees and attempts to accommodate them were thoroughly compromised from the beginning. This subversion created a baffling problem the GVN could not resolve and soon contributed to the general de-stabilization of both Pleiku and Khontum provinces.

To complicate matters and exacerbate ill feeling the GVN then had a go at resettling the Montagnards themselves. Abuses abounded. GVN troops forced many Montagnards out of the hills at gun point ostensibly to remove them from the clutches of the Viet Cong. They were separated from their ancestral lands and sites of the graves of their ancestors in an inexplicably cruel manner.[25] Once resettled the Montagnards typically were given land too poor for cultivation and very little funding. The lion's share of the development budget went to the Vietnamese settlers—visibly so.

The crowning blows to Montagnard quietude though were delivered soon afterwards by the artillery parks created around Pleiku City. The guns routinely fired away every evening at smoke from Montagnard campfires in the distant jungle. The hill people were trying to cook their evening meal over these fires and in most instances they had done nothing to warrant such grotesque maltreatment. However, the rationale behind this indiscriminate shelling was that if the fires were more than two or three kilometers away the people who lit them were probably infiltrated by the communists. Not exactly true at the time, the underlying assumption was nonetheless prophetic.

Slowly but surely sufficient discontent began to build up among the Montagnard for the Viet Cong to make some notable inroads among certain groups. The communist ploy essentially involved playing upon their fear of losing their land and traditional way of life.[26] Heaps of GVN abuse made this an easy sell since by this point almost anything would have had more promise than GVN fare. Subsequent bids by CIA operatives and United States military personnel to remedy the situation spiraled out of control and produced a no-win scenario that adversely affected the security of the entire plateau region. United States projection of large conventional forces into the area ultimately attracted a substantial GVN counter force and this led to the interspersion of multiple counterinsurgency type skirmishes with conventional set piece battles involving Viet Cong, North Vietnamese Army (NVA), United States, ARVN and other United States allied troops.

Laymen tend to ascribe an irregular warfare cachet to this mixed bag of confrontations. However, as with the fighting in other parts of Vietnam the highland clashes by and large were not true guerilla warfare actions. Many of the forces in play were not authentic Special Forces. Certainly, quite a few of the men involved were elite troops by any standard but by the same token they were not pure guerilla fighters in the usual sense of the term. They made too much use of technology and their utter dependence upon helicopters turned into a crutch instead of a tool.[27] Hence, these forces were engaged in an exaggerated and perverted form of what the British casually refer to as jungle bashing.[28]

The British also made a few blunders in Malaya where they were not the perfect purveyors of low-intensity warfare despite the prowess of some of their men. However, they quickly learned from their mistakes and resisted the temptation to throw more men and material at the problem in the hopes of winning by attrition. Instead they chose to learn from experience, to emphasize quality over quantity and to perfect their operations. The United States departed from British experiences on this score and sought victory through firepower and escalation sustained by the rotation of large numbers of men in and out of the battle zone. Whereas the better stratagem would have been to leave fewer men in harms way for much longer periods of time. Ultimately, these machinations only sullied the waters in which the Senoi Praaq had to swim and created a larger problem of such magnitude and scope it soon spiraled out of control.

THE SENOI PRAAQ MODUS OPERANDI AND PAT NOONE'S GROUP PSYCHOLOGY

The DRVN, in shared complicity with the GVN and the United States, probably never intended to honor the commitments it made in Geneva. The consolidation of the two Vietnam's into a unified communist state became Hanoi's principal concern upon partition of the country. To accomplish this goal South Vietnam had to be invested from both within and without by substantial DRV forces. Naturally, not long after the accords were signed men and materiel began to move south in considerable numbers. These combatants had to skirt Montagnard areas. The Diem government was aware of the danger they posed and this is where the employ of the Senoi Praaq came into play. By the time the initial Senoi Praaq troopers arrived in March 1963 only a fraction of the overall Montagnard population had been corralled into indigenous resettlement centers. The remainder lived in their traditional long houses, built upon stilts, throughout Pleiku province and they could if properly approached be used to advance GVN purposes.

These traditional settlements (sometimes the construction was substantial and semi-permanent) often functioned as temporary map coordinates without names to identify them for Senoi Praaq troopers on the ground. Montagnard practice of slash and burn agriculture obliged many of them to shift from one swidden to the next every few years. Senoi Praaq mobility and their compatibility with the Montagnards put them in a position to do what the GVN could not do for itself. Thus, Senoi Praaq emissaries who canvassed these settlements kept constantly on the move networking from one group to the next. Typically they brought token gifts with them as was their own custom and made arrangements for the provision of small arms which the Montagnards might use to hunt and defend themselves. Firearms were a popular item—shotguns in particular. More often than not—at least initially—the tribal elders subsequent to Senoi Praaq overtures chose to align themselves with the GVN.

Generally speaking, to subvert a group of individuals one or more of their own kind must be used to betray or manipulate the others. Making the initial approach to one's prospective target can be difficult particularly in those instances in which ethnic boundaries must be crossed. No such barrier existed for the Senoi Praaq operating in Vietnam among the Montagnards. The key to securing Montagnard cooperation lay usually with the village council's willingness to go along on a strictly voluntary basis with Senoi Praaq proposals. Grass roots information secured in this fashion would then be employed collectively to generate strategic intelligence about overall Viet Cong activities in Pleiku and parts of Khontum province.

Their plan was sound and in principle easy to implement but it was only with some difficulty that Senoi Praaq emissaries were able to proceed in this manner. Montagnard society—like Orang Asli society—was group oriented and much more egalitarian than most. It was not simply a matter of buying off the leader and using him or her to control the others. Each member of the tribal council or inner leadership core had to be convinced of the wisdom and necessity for a particular course of action. In the instance of a split vote or dissension the group tended to either take no action or hive off into new clusters of extended family units.

For their part neither the council members nor tribal elders ever took part in a decision that threatened the group. Pat Noone made this key discovery during his earlier field work among the Orang Asli when he noticed that various headmen had power but concluded that there were multiple checks in place to this authority.[29] Life in the jungle was much more difficult than most might imagine and the survival of any one individual was contingent upon the overall well-being of the entire group. Their environment saw to it that the Montagnards and Orang Asli alike evolved in such a manner that aberrant non-conformist behavior was seemingly bred out of their gene pools. The instinct for collective survival in each individual involved in the decision making process had to be satisfied before the council threw in with an idea or proposal. An offer to guarantee the security of the group was a necessary part of any proposal put to any Montagnard.

Thus, every approach took time and once a relationship was established it had to be cultivated continuously.[30] But despite these many necessary and intricate requirements for making a successful approach, Senoi Praaq troopers were able to propagandize convincingly, make some headway amongst the Montagnards and marshal a number of them for training. It was not long before they built up a promising force of loosely controlled tribal allies.

CIDG forces in general had a substantial initial impact upon communist activities in the highlands especially in Khontum where the Mountain Scouts *per se* were credited with driving the Viet Cong from the province. CIA sources with access to captured enemy documents even reported a drop in Viet Cong force strength in this particular area from 5,000 to 500 men between January and March 1962.[31] Unfortunately, too many cooks spoil the broth and Senoi Praaq success within the greater scheme of things would ultimately be short-lived.

Chapter 14

The Situation Begins to Unravel

While the opportunity to participate in the BRIAM venture had been a godsend for the Senoi Praaq, Vietnam came to be a most unhappy chapter in its history. An early exit helped the unit avoid the stigma associated with the disaster which ultimately befell the country but not without its troopers bearing witness to the makings of a sobering debacle. Vietnam developed into a focal point of superpower attentions where the pattern of engagement was the inverse of that in Malaya.

FEW INSURGENCY SCENARIOS ARE EVER ALIKE

The Malayan Emergency started with the deployment of fairly large MRLA combat units that gradually decreased in size as the conflict played out over time in true style of guerilla warfare. British and Malayan security forces managed to break communist supply lines through their successful implementation of the Briggs plan and their destruction of the Min Yuen organization operating in Malaya's urban communities. They also quickly abandoned attempts to pin down the Orang Asli in resettlement camps and concentrated instead upon winning their hearts and minds. This in turn compelled main force communist units to break up into fairly small groups who then were obliged to spend a great deal of their day foraging for food and developing their own intelligence. For a brief period they managed to compensate for this setback by clever use of the Orang Asli but soon lost much of their initial tribal support. Their integrity was thereby degraded and their capacity for combat denigrated. At this juncture the communist guerilla fighters became easy targets for the Senoi Praaq troopers who ran them down over a period of time with little difficulty.

The pattern of military activity was quite different in Vietnam where a limited covert struggle initiated by Viet Cong guerillas escalated into a modern conventional war. In the crucial central highlands region—and throughout the RVN—hurried implementation of the strategic hamlet program left little time for GVN to learn from its mistakes. This hasty, bungling response contrasted sharply with the measured, slow-paced counterinsurgency measures undertaken in Malaya. Moreover, the strategy's brutish application among the Montagnards came back like a boomerang to thwart the strategic hamlet initiative in the highlands. Robert Thompson's concept of "reaction forces" failed too. The reaction forces brought in to support the hamlets quickly turned into grossly overblown and overplayed airmobile operations. None of these efforts cut off DRVN supply lines via the Ho Chi Minh trail which snaked ever so irresistibly through Laos, Cambodia and South Vietnam. The communist forces actually drew strength from these GVN and United States countermeasures, and this is how the war escalated into unfavorable terms. This loss of the highlands ultimately provoked the collapse of the GVN, bringing a dramatic end to this long and dreadful struggle.

As this unsettling trend line took shape in the general evolution of the war several salient points germane to the Senoi Praaq experience became clear and in retrospect these conditioning factors are easy to identify. The counterinsurgency program as it was applied in the highlands came to be slanted towards a spectrum of covert activities whose objectives extended far beyond the limited mobilization of the Montagnards; its civilian emphasis ultimately gave way to a military focus; and high technology solutions supplanted the low technology approach followed in Malaya. All these directional changes transpired in an environment partial to Vietnamese rather than Montagnard interests. This bias of many if not most United States officers in favor of the Vietnamese interfered with their impartial judgment about the Montagnards and it diminished the effectiveness of all counterinsurgency activities in the central highlands.

BIAS TOWARDS THE VIETNAMESE

The United States war effort in the highlands took place within the context of country-wide engagement (i.e., South Vietnam) in which the Vietnamese were the paramount and politically dominant ethnic group. United States policy thus had to be weighted in favor of Vietnamese, not Montagnard, concerns even though United States interests in the Pleiku region itself lay in cultivation of the Montagnards. Or, so American policy planners thought. This mental leaning in favor of the Vietnamese became all too clear to the Montagnards and over the long term it undermined the integrity of American operations on the ground and sowed the seeds of their faint cooperation and in some instances outright defection.

Conversely, the British in multi-racial Malaya were under no such constraints, self imposed or otherwise. They had great leeway in their dealings with the Orang Asli tribes. The Malay's were the politically dominant native group but they were not overlords. The British had imported so many Chinese and Indians—to work Malaya's fabled tin mines and rubber plantations—their sheer number put great pressure on the Malays to be ethnically objective. As a consequence the British/Malay alliance could easily solicit Orang Asli loyalty and play the tribal minority card in Malaya, just as the French had done earlier in Indochina through their own favoritism towards the Montagnards.

The French for their part employed a divide and rule strategy in order to take advantage of intense regional rivalries among the Vietnamese throughout the whole of Vietnam (i.e., North and South Vietnam). Thus, they could deal freely with the Montagnards as a distinct ethnic entity. Indeed, the French administration left many highland areas undisturbed by any developmental activity. They were akin to autonomous territories duly recognized by the government of France in the post-World War II Declaration of May 27, 1946. The Montagnards in these areas enjoyed a high degree of self-determination and they served willingly as an effective buffer for French interests. French pragmatism and objectivity had conditioned the Montagnards to expect a more even-handed treatment from the Americans than they ultimately received.[32]

Instead, relatively little of the United States economic development aid donated to Vietnam ever reached the central highlands. Whatever aid there was for Montagnard programs per se was channeled through the GVN and designed to foster Montagnard assimilation into a "Vietnamese" South Vietnam.[33] These programs, bringing modern agricultural practices and a static life style ill-suited for semi-nomadic hunter gathers, quickly threatened their traditional culture with extinction and provoked great resentment among the Montagnard peoples. Whereas Malaya became "multiethnic" Malaysia subsequent to independence the GVN never offered its own ethnic minority peoples such an umbrella of protection. GVN belatedly established in the late 1960s a Ministry for Ethnic Minority Development along DOA lines in Malaya. Once again, though the idea was sound in principle and had precedent in Malaya it did not accomplish much upon application in Vietnam. The Ministry by and large was an empty gesture too late either to repair the damage done or to improve GVN-Montagnard relations.[34] Thus, GVN failed to evolve a political structure capable of channeling the energies of its minority peoples in the fight against communism.

Independent SVN quickly developed into a client state of the United States. Its government beset by internal political squabbles never managed to address forthrightly the indigenous issue. Some Montagards nonetheless went on to fight alongside United States forces but as individual mercenary groups pursuing narrow self-interest rather than as allies with a real stake in the war's outcome. Montagnard populist energies did coalesce anew and a small independence movement, the United Front for the Liberation of Oppressed Peoples (FULRO) was formed but it never made much headway with the GVN. This organization had its roots in the BaJaRhaKo movement but it was unstable. In point of fact, mercenary forces built up by the United States and loyal to FULRO mutinied briefly in mid-1964 killing some eighty South Vietnamese troops around Ban Me Thuot. However, their American advisors diffused the situation and the GVN made a few concessions but nothing of consequence really changed.[35] Thus, the Montagnards were never properly exploited and ultimately the GVN aborted its effort to control the highlands and as everyone well knows collapsed not long thereafter.[36]

DEFILEMENT OF THE SENOI PRAAQ MISSION

The BRIAM initiative in Pleiku province, albeit novel, was only one of many security development projects undertaken early in the war. Nearly all these programs were geared towards the defense of the ethnic Vietnamese population in various parts of South Vietnam. A few projects targeting North Vietnam proper involved CIA sponsored raids along the northern shores near the port of Hai Phong and airborne "insertion missions" to drop South Vietnamese commandoes. These failed missions attempted to sabotage commercial/military targets and to establish a network of agents. Such aggressive tactics in play early in the war were spurred on some say by pique over the debacle in Cuba at Playa Giron.

However, none of these activities were successful in taking the war to the enemy and did little if anything to weaken communist resolve. Viet Cong casualties may well have been upwards of fifty percent of their force strength in any given year but initially this did not deter DRVN. The progressive expansion of communist influence over the South Vietnamese population more than compensated for these losses. The Viet Cong apparatus would not overreach itself until 1968 subsequent to the Tet Offensive and several follow up actions. In the interim United States estimates of Viet Cong strength continued to be revised upwards from approximately three thousand fighters in 1959 to 15,000 by mid-1961.[37]

These early counterinsurgency efforts though broad in scope lacked both coherence and adequate depth in any particular applications area. They were disjointed and their profusion diluted the effectiveness of Senoi Praaq activities in the highlands at the very time they were most needed. The visible focus of Viet Cong and NVA attentions, for that matter, had begun to shift from Saigon and the Mekong River Delta regions to the central highlands. Viet Cong forces in these low land areas engaged in the struggle to destabilize South Vietnam had been badly mauled in 1960-1961, and the DRVN needed new venues in order to maintain an offensive posture. The highlands were remote and difficult for RVN and United States forces to access; thus they held out the tantalizing prospect of sanctuary for communist fighters. If they could capitalize upon Montagnard restlessness and dominate this region, abutting the Ho Chi Minh trail, it might then serve as a secure base area from which to menace the remainder of South Vietnam.[38]

THE MILITARY TAKES CONTROL OF THE COUNTERINSURGENCY EFFORT

Lackluster performance by civilian agents and police advisory personnel employed by Michigan State University early in the war undoubtedly explains to some extent why responsibility for the counterinsurgency program in South Vietnam and the covert war against North Vietnam was turned over to the regular military. There was also concern in Washington that the CIA not be compromised by another fiasco upon the heels of its abortive activity in Cuba.[39] This transfer of control to the regular military turned out to be a wise move given that the Village

Defense Program/Strategic Hamlet Program became one of the worst high profile project failures of the war. The original intent of the program as stated/implemented by the CIA—according to former agency representatives—had supposedly been more or less defensive in nature.[40]

A reorganization of United States efforts and an expanded role for the regular military entailed the dispatch of more elite forces to Vietnam and a buildup in the Pleiku region. Again, this ran counter to British prescriptions for management of the conflict which emphasized paramilitary type units and a strong civil police role. Even though they continued to give some attention to the use of special forces operations the regular military called the shots and introduced so many conventional troops into the fighting the overall ratio of special operations and special operations type units to conventional forces never really increased. If anything, it declined in the overall scheme of things even as more and more Montagnards were levied to fight in this or that capacity against Viet Cong and NVA forces.[41]

Thus, the Senoi Praaq could do little more for the Montagnards once the war entered the phase of American escalation. Their objectives differed from those of the Americans in that the Senoi Praaq contingent's role in the Central Highlands was largely designed to be passive and unobtrusive, not just defensive. BRIAM proper for its part functioned as an advisory body and it sought to value maximize through more efficient application of local resources in the conflict.

LOW TECHNOLOGY DE-EMPHASIZED

And so, by 1961-1962 as the overall American posture in South Vietnam's highlands tilted firmly in favor of the Vietnamese all too often to the detriment of the Montagnards, United States forces began to build up on the ground. First came a plethora of special forces advisors and ultimately the Marine landing at Da Nang in 1965. The escalation was brought about in large part by an inability of existing assets to protect United States airfields around Pleiku city. Heavy Viet Cong attacks later in 1964 made it abundantly clear that the area had become unstable and insecure. Invariably a corresponding methodological shift occurred as a sequitur to the deepening United States involvement. The new focus of the intelligence game was high—as opposed to low—tech solutions for intelligence gathering. The Senoi Praaq had emphasized human intelligence exclusively and theirs was definitely a low-tech approach.

Rather than faithfully emulate the British strategy in Malaya the Americans opted to follow the course of action more or less taken by the French some years earlier in their ill fated struggle against the Viet Minh which placed great emphasis upon mechanical technology, radio intercepts and code breaking in addition to human relations. Signals intelligence supplemented by reconnaissance imagery, data acquisition through sensors, and a panoply of gadgets became the order of the day. British officials may have been inclined towards chartism as a means to suppress the Malayan Emergency but the Americans in Vietnam were real number crunchers. Pentagon officials and policy planners at the White House required incredible quantities of data to make war in SVN and engaged in a titillating sort of "speculative necrology" to monitor the progress of the war.[42] Unfortunately for both the French and the Americans the NVA and Viet Cong were, for all their acumen as fighters, decidedly an old fashioned low tech enemy when it came to communications.

The North Vietnamese tended to use couriers in SVN as did the MRLA in Malaya and there was not a great deal of radio communication to intercept. Broadcast instructions per se were relayed via courier and had little meaning to anyone but their individual recipients. DRVN leaders knew they were technologically deficient and observed elaborate procedures to keep to a minimum their exposure to United States Communications Intelligence (COMINT).[43] The American predilection for high-tech intelligence gathering is indeed ironic in so much as the Senoi Praaq had devised elaborate but simple schemes to snare couriers and to penetrate or otherwise break up agent networks. These sophisticated techniques as they were applied particularly in the central highlands yielded a dearth of real targets for the American forces. Consequently, they more or less squandered their overwhelming firepower—albeit devastating when accurately targeted and indeed there were many deaths—against an elusive enemy. Once total tonnage of expended munitions is juxtaposed against North Vietnamese population statistics the failure of this intelligence strategy in which liberal use of firepower played a cornerstone role becomes all too apparent. Thus, very little information of substance ever made it into the hands of the Americans and what little intelligence they did pick up proved to be of inappreciable value in operations.[44]

In this regard the United States dropped four times more bombs on Vietnam than it did throughout the whole of the Second World War. The tonnage involved accounted for no small part of the war's enormous price tag and its impact upon the battle field was much the same as the horrendous bombardments on the western front by German and allied forces during the First World War. Just as in Flanders, or at the Somme, the ground in many parts of the Vietnam became so battered and pockmarked it resembled a lunar landscape.[45]

Thus, not only did it become difficult to operate in such areas with any degree of efficiency but the people themselves were lost as a resource.[46] Communism like all ideologies or religious appeals feeds upon human misery and thrives amidst lost souls. And so, just when the Montagnards needed quality help the most, the United States forces created all the requisite conditions whereby they would deny themselves the help of the Senoi Praaq. This defilement of the Senoi Praaq mission and the failure of subsequent high technology counterinsurgency opened a gateway for the DRVN army to push southwards into Saigon.

Part IV: Notes

1. Clearly, Robert Thompson was very much involved in the disposition of GVN forces and operated very closely with key ARVN personages. These observations and information about forces available in the IV Corps region are extracted from correspondence between Robert Thompson and F. A. Warner of the Foreign Office. The letter in question also concerns a meeting between Robert Thompson and Ambassador Lodge. Both the cover letter and the details of the Thompson/Lodge meeting are expanded upon in subsequent footnotes. See Thompson to Warner, 18 Sept. 1963, PRO, BRIAM (i.e., FO) Vietnam/371/170102.

2. This figure circulated among top American officials and may be found in an outgoing telegram from Saigon authorized and signed by Irwin M. Tobin. The transmission is dated 6/26/61. See, Department of State, Central Files, 751K.00/6-2661, National Archives, Washington, D.C. However, Roy Jumper Sr., the first American political science scholar to do research in independent South Vietnam suspects these numbers were much higher and bases this observation on personal experiences in Vietnam during the period 1954-1962. Prior to Geneva most villages in the SVN delta region were believed by people nearest the action to be under Viet Minh influence. For a provocative and original analysis of the communists in SVN, their attraction for the peasants, and the difficulties encountered by the Diem regime in countering their appeal and rooting out their influence see Roy Jumper, "The Communist Challenge in South Vietnam," *Far Eastern Survey* XXV (November 1956): 161-168.

3. Clutterbuck's treatment of the Malayan Emergency—in which some contrast between Malaya and Vietnam is provided—makes available similar numbers. Sir Robert Thompson's forward to this work lists corresponding figures for the Vietcong at 5,000 guerillas and 100,000 supporters. Sir Robert also agrees things might have worked out for the better— at least in the early stages—had the Malayan formula " . . . been suitably adapted and consistently and intelligently applied." See Richard Clutterbuck, *The Long Long War: The Emergency in Malaya 1948-1960*, with a forward by Sir Robert Thompson (London: Cassell, 1967), viii.

4. The BRIAM/British Embassy position vis prospects for a successful resolution of the war was stratified. British consular officials tended to be somewhat more pessimistic than BRIAM personnel. Their differences come to light in correspondence between H.A.F. Hohler and E.H. Peck. See Hohler to Peck, 20 Mar. 1963, PRO, FO Vietnam/371/170100.

5. This figure is cited in a letter *cum* report exchanged between H. A. F. Hohler and

Lord Home dated January 30, 1963. Ambassador Hohler's insights are very revealing and are drawn upon extensively by this work. See Hohler to Home, 30 Jan. 1963, PRO, FO Vietnam/ 371/170100.

6. Actually, Sir Robert is known to have made the Kuala Lumpur/Saigon shuttle several times prior to this point. He soon became a major strategist in the conduct of the war and GVN officials took more stock in his advice than the Americans. Be this as it may, Douglas Valentine makes several interesting references, in his treatment of the Vietnam experience, to Robert Thompson and the application of police/intelligence procedures. Thompson is also said by Valentine to have been lined up initially by Roger Hilsman, director of the State Department's Office of Research and Intelligence. See Douglas Valentine, *The Phoenix Program* (New York: William Morrow and Company, Inc., 1990), 75, 79, 80, 92.

7. The trips/discussions are reported by J. A. Menderhall in a telegram from the American embassy in Vietnam to headquarters, Washington. The transmission is dated October 29, 1960. See, Department of State, Central Files, 751K.00/10-2960, National Archives, Washington, D.C.

8. A number of American authors make fleeting reference to the Mountain Scout and other reconnaissance oriented programs seemingly unaware of the Senoi Praaq link to the covert insurgency program in Vietnam. For some of the more recent examples see Jerald Cannon Hickey, *Free in the Forest: Ethnology of the Vietnamese Highlands 1954-1976* (New Haven: Yale University Press,1982), 76, 77, 78, 87. John Prados, *The Hidden History of the Vietnam War* (Chicago: Ivan R. Dee, 1995), 72,75,79. Thus, few if any written references to a specific Senoi Praaq role in Vietnam exist to date other than classified British records. Those found in SEATO files remain sealed to present. Protocol governing international organizations is such that each of the member states must agree to declassify documents before material can be released. And such agreement rarely occurs when numerous member states are involved as in the SEATO case. There are undoubtedly too many skeletons in this closet for their transparency to ever become palatable.

9. British police personnel tended as we know to work solo in small Malay villages and among the Chinese. Although they did have some contact with the Orang Asli at this early stage, the Orang Asli were not their main focus.

10. To their credit the embassy people did occasionally send their representatives on casual observational tours and a number of noodle-doodle reports that stem from these forays, into the wilderness and away from the usual comforts, are in the archives. Invariably, they are fairly detailed but the report writers often do not see the forest for the trees let alone a forest of tigers. In this respect, Tiger Beer was very popular in Southeast Asia and could easily have reached Vietnam's central highlands through R.O.D. Noone's supply train in the form of a gift. This would have been a good way to break the ice as Montagnards love to drink. A traditional greeting among them is "Nam Loy" i.e., "come let's get drunk" (together). The reporter in this instance, T.J.C. Heavner, operating out of the old imperial capital at Hue, should have viewed the very presence of commercial beer in large quantities among such poor people as more than a matter of simple curiosity. Be this as it may, his observations suggest R.O.D. Noone may have been in Pleiku well before the initial meetings between British and Vietnamese VIP's. See, Department of State, Central Files, 751K.00/1-560, National Archives, Washington, D.C.

11. They were also at odds with the string of Saigon regimes put into place by the Americans upon French withdrawal from and partition of Vietnam. For a revealing inquiry into the Cao Dai and their role in Vietnamese political processes see Roy Jumper, "The Cao Dai of Tay Ninh: The Politics of a Political-Religious Sect in South Vietnam," in Balkrishna G. Gokhale, ed., *Asian Studies: A Collection of Papers on Aspects of Asian History and*

Civilization (Bombay: Popular Prakashan, 1966), 142-154.

12. Australia sent troops to Vietnam in May and June of 1965 in batches of 800 and 400 respectively. They were preceded by about 80 advisors who were already on the ground at the time. As of 1963 the Australians were headed by Colonel Ted Serong, leader of the Australian Training mission to South Vietnam. His focus tended to be upon the delta and the area around Saigon. The strategic hamlet program in these areas was designed to destroy the Viet Cong whereas in the highlands their idea was to restrict enemy movement. See, United States Department of State, *Foreign Relations of the United States, 1961-1963*, Volume 3, Vietnam, January-August 1963, (Washington, D.C.: GPO, 1991), 315- 316. Curiously enough though he is also reported to have worked with the Montagnards; to possess an intelligence background acquired during the Malayan Emergency through experience with the Police Field Forces; and to have built himself a villa with money from USAID outside of Dalat. See Valentine, 92, 94, 231. Sarong spelt with an "a" as opposed to an "e" incidently is the term of reference in Bahasa for a garment worn around the lower half of the body by Malays and some Orang Asli in Malaysia.

13. Senoi Praaq sources named names and finger Cols. A. (US CIA), M. (US Special Forces), S. (US Army) and McC. (US Army) as the key American personalities who interacted with R.O.D. Noone at the ground level. R.O.D. Noone himself held no official rank except the courtesy title of Colonel extended to him by US personnel.

14. Eventually, the mountain commando/trail watcher programs became the mountain scouts/border surveillance units which were in turn rolled into the CIDG program upon initiation of Operation Switchback. Consolidation reportedly took place in the Spring of 1962. See Francis J. Kelly, *U.S. Army Special Forces: 1961-1971* (Washington, D.C.: Department of the Army, 1973), 32, 33.

15. For a time President Diem and his brother Nhu personally directed these programs out of the Presidential palace with Robert Thompson's input and advice. However, United States military initially reacted angrily to Thompson's proposals which ran counter to their conventional strategy for dealing with the Communist challenge to South Vietnam. David Kaiser, *American Tragedy: Kennedy, Johnson and the Origins of the Vietnam War*. (Cambridge: Harvard University Press, 2000), 167.

16. Veteran British journalists, John Bierman and Colin Smith offer a telling account of Wingate's story. Sir Robert (Bobbie) Thompson was, for his part, a Royal Air Force (RAF) liaison officer at the time and took an active part in both Chindit operations. See *Fire in the Night: Wingate of Burma, Ethiopia, and Zion* (New York: Random House, 1999), 293, 387.

17. However, Noone did have at least one face to face meeting with President Diem to discuss the highlands and the Vietnam situation generally. The encounter was a memorable one for Noone as President Diem spoke for two hours straight. However, Noone was reported to have told a friend sometime later that he felt it was impossible for him to converse with Diem—such was the man's propensity for endless monologue.

18. For that matter, the reference was made posthumously and in tribute. See Robert Thompson, *Peace Is Not At Hand* (London: Chatto and Windus, 1974), 4. Thompson is also on record in a memorandum of conversation as saying that they were fickle or "prone to switching sides" and had to be handled very carefully. See, United States Department of State, *Foreign Relations of the United States, 1961-1963*, Volume 3, Vietnam, January-August 1963, (Washington, D.C.:GPO, 1991), 194.

19. The Jarai speak a form of archaic Malay. Some of them also spoke French and this foreign language ability further enhanced the possibility of their communication with British officers. Most British officers typically have some French language training of their own.

20. The Semlai reside towards the south of the Malayan main range and are identified as a Proto Malay or Jakun type peoples.

21. Although the Senoi Praaq had free run of both provinces, and are also reported to have made forays into Cambodia, they tended to focus upon Pleiku.

22. The LLDB detachment was led by a Vietnamese lieutenant called Muy.

23. Lt. Muy was a curiously diminutive but authoritative little creature and he became R.O.D. Noone's bete noir in this regard.

24. R.O.D. Noone, in addition to being an effective communicator, spoke excellent French and this helped a lot in his relationships with the Vietnamese Army and General Khanh who is said to have been a good soldier. However, neither attribute would turn out to be very effective in abating Vietnamese abuse of the Montagnards. Latent hatred between the two groups was too great to overcome and to further complicate matters the LLDB was run out of the palace by President Diem and obviously there were limits to R.O.D. Noone's influence.

25. Truong Nhu Tang, *Journal of a Vietcong* (London: Jonathan Cape, 1986), 46-47. From Truong's account we also see that Diem's interest in the Strategic Hamlet Program dates to 1957 when he sent Albert Thao (Pham Ngoc Thao) to Malaya. This led to the Agrovilles which fizzled out in a year or so only to be resurrected through the "village defense" scheme equally as flawed as the disastrous Strategic Hamlet Program. The pathology of the Strategic Hamlet Program is well known but in this unique account of its history much credit for its failed implementation was given to an undercover communist agent who wielded great influence deep within the GVN.

26. David Nuttle of International Voluntary Services first brought the swell of Viet Cong activity among the Montagnards to the attention of United States authorities. This led to a period of continuous contact between CIA/special forces operatives and the highland peoples. Nuttle's original report is available as an attachment to a dispatch issued by J. A. Menderhall from the Embassy in Saigon to the Department of State, Washington,D. C. dated October 28, 1969. See, Department of State, Central Files, 751K.00/10-961, National Archives, Washington, D.C.

27. True, helicopters enable a commander to multiply his presence at will and they are a must for certain tasks. However, they drink gasoline as payment for these services. This requires gasoline stockpiles be accumulated and the depots that are needed to store the gasoline must themselves be well-guarded less they become easy targets for mortar attack. And so obviously helicopters induce escalation and present one's enemies with opportunities.

28. In British military vernacular jungle bashing means patrolling the jungle, aggressively at times perhaps, but not bashing it literally as during implementation of the more lurid search and destroy concept. Search and Destroy actually embodies a type of warfare in which an overabundance of force is committed and "means" are not in sync with "ends" *vis-à-vis* liquidation of one's enemy.

29. The Orang Asli speak various languages/dialects and Malay is their *lingua franca* but the term Tok Batin is generally understood to refer to an important chief.

30. One could never afford to be caught in a lie or make promises and then not deliver. Plutark himself was probably not much better at recognizing flatterers and their false promises than the Montagnards. That the Montagnards were upon conclusion of the war, some ten years after the Senoi Praaq left, hung out to dry can only be attributed to one reason. The United States special forces advisors sent out to win their trust (and to whom the Senoi Praaq had lent their own credibility) were lied to as well. United States government bias in favor of the Vietnamese prevented any genuine commitment towards the highland

peoples and this favoritism continued even on the eve of defeat. Very few Montagnards were evacuated to the United States upon GVN's collapse. In all fewer than a thousand are reported to have made good their escape and this estimate is inclusive of those who got out at the wars end. The American betrayal, at the Department of State's behest, of the Montagnards in South Vietnam is also reported to have surpassed that of the Hmong allies in Laos. See Prados, *Hidden History*, 86.

31. Hickey, 78.

32. By the time the United States became involved there was already a Montagnard independence movement in place called the BaJaRhaKo. It seized and briefly held Pleiku in 1958. This act should have been a warning signal to those who sought to dabble in tribal manipulation. Nonetheless American policy makers targeted the highlands where they actually worked at cross-purposes *vis-à-vis* Montagnard interests. They promoted Senoi Praaq activities among the Montagnards at the same time they were undermining their interests through generous patronage of GVN. True, they had at least to give the appearance of catering to GVN whims over the Montagnard issue given that the South Vietnamese were in both an ethnical and political sense the preponderant people. However, as a practical matter United States personnel were dealing with only a half-loaf in Vietnam and could have chosen to play at more than *ramasse-miettes*. Their undertakings in the highlands consequently had a schizophrenic-like quality about them largely absent in British dealings with the Orang Asli in Malaya. Their brash, sometimes naive, actions also contrasted observably with the *savoir faire* and the diplomatic finesse (and sneaky behavior when the occasion permitted) exhibited by the French.

33. The tenants of realism to the American way of thinking dictate deals be made with representatives of a sovereign nation state to the extent this is possible. Hence, it comes as no surprise that United States Department of State representatives peddled the view that the Montagnards would have to assimilate with the Vietnamese. T.J.C. Heavner's own views, in the aforementioned provincial report concerning the Montagnards, are extremely biased in this respect and it is obvious that he looks down on the Montagnards as an inferior people. Hence, just as the significance of Tiger Beer in the Highlands is overlooked so to is the impractical nature of dealing with the Montagnards as though they were candidates for assimilation—"...the Consulate believes the Montagnards can be led peacefully if not happily into the 20th century." See, Department of State, Central Files, 751K.00/1-560, National Archives, Washington, D.C.

34. Hence, the moribund General Directorate for Montagnard Affairs became the Ministry for Ethnic Minority Development only in 1967. See Prados, *Hidden History*, 83.

35. Prados, *Hidden History*, 80-81.

36. Prados, *Hidden History*, 85. For a gritty personal account of one Special Forces officer who in defiance of U.S. policy joined forces with the FULRO leadership in an ill fated attempt to establish an autonomous free territory for the Montagnards see Don Bendell, *Crossbow* (New York: Berkley Publishing Group, 1990).

37. Estimates of enemy manpower may or may not have been low but otherwise the analysis of Viet Cong intentions was fairly accurate. These observations are made in an outgoing airgram signed and approved by Sterling J. Cottrell, dated Sept. 2, 1961. See, Department of State, Central Files, 751K.00/9-561, National Archives, Washington, D.C.

38. More specifically, the areas in question were said to be adjacent to the Lao Frontier where the Viet Cong were thought to plan the formation of an offensive enclave. This information appears in the aforementioned outgoing telegram of June 26, 1961 from Saigon authorized and signed by Irwin M. Tobin. See, Department of State, Central Files, 751K.00/6-2661, National Archives, Washington, D.C.

39. Hickey, 80.

40. Here Gerald Hickey quotes Lucien Conein. See Hickey, 80-81.

41. The downside to military control over special operations is fairly obvious. There is an inbred distrust among members of the regular military for special forces personnel and their programs. As a general rule military commanders are wary of the special forces and consider their complement as little more than unbalanced and un-trustworthy killers. David Hackworth further makes the point in his memoirs that special forces type units within the United States military *per se* are frequently betrayed. Obviously the relationship between commands is not a healthy one. Given such institutional proclivities the Montagnards invariably faired badly once they became mixed up with special forces run by the United States military. For an interesting summary of United States failures in Vietnam see David Hackworth and Julie Sherman, *About Face: The Odyssey of an American Warrior* (New York: Simon & Schuster, 1989).

42. This catchy locution appears in several works but the author is unable to ascertain who initially coined the term. However, it sounds like Robert Thompson's style.

43. North Vietnamese intelligence kept radio traffic to a minimum. It broadcast a limited number of cryptic messages one way from Hanoi to a few receivers in the south at certain dates and times. The recipients then forwarded these messages via courier to agents who knew the meaning in advance. Prados, *Hidden History*, 195. They may also have been using some sort of rudimentary one-time pad system.

44. However, there were notable exceptions in which "lucrative targets" were acquired. At times and in certain areas targets could not help but be concentrated, such as along the Ho Chi Min Trail. John Prados, *The Blood Road: The Ho Chi Minh Trail and the Vietnam War* (New York: John Wiley & Sons, Inc.), 195-196. Clearly, as the war progressed not only did more traffic become available for interdiction but North Vietnamese communications requirements became more extensive and thus more vulnerable to intercept.

45. The author, one of the first Americans to return to Vietnam, witnessed the lasting effects of this pounding. Nearly fourteen years after the wars end in 1975 the terrain across much of the country remained badly scarred. For a classic account of the futility associated with this type practice see *In Flanders Fields* by Leon Wolf. The misguided application of brute force is an underlying theme in a classic book otherwise devoted to intransigence and utter, pointless disaster.

46. John P. Cross, in *Jungle Warfare: Experiences and Encounters*, is adamant about the need to avoid mismanagement (i.e., be efficient) and escalation in CRW lest matters get out of hand. He reinforces these points and further adds in *A Face Like A Chicken's Backside: An Unconventional Soldier in South East Asia, 1948-1971,* that large-scale encounters never seemed to work very well either. See Cross, *Experiences*, 175, 178 and *A Face*, 30.

PART V

The Senoi Praaq Stands Down in South Vietnam as the Indonesian Crisis Erupts

The Vietnam situation had indeed begun to go bad by early 1963 and not long after their deployment the Senoi Praaq position in the central highlands was throughly compromised. Just about every step taken to contain or suppress communist revolutionary warfare had made matters worse instead of better. As a consequence R.O.D. Noone had begun to circulate word informally among his field commanders that the outlook appeared hopeless.

Some of the first in were soon to become the very first out and to relinquish their operations to United States special forces altogether. Provisions underway in Malaysia for dispatch of additional Senoi Praaq elements inclusive of the organization's key logistics person—Ahmad bin Khamis—were abruptly canceled. There would be no second wave of Senoi Praaq to descend upon Pleiku or any further effort to exploit directly the Montagnard situation.

Matters in general were also about to get much worse for the British at the national and international levels. Word of American bungling of their prescriptions for the war would reach into the presidential palace itself and compound the leadership deficiencies from which GVN already suffered. President Diem of South Vietnam (SVN) himself would be assassinated by the end of the year and Sir Robert's man in Saigon would be no more.

To complicate things even further, back at home problems with Indonesia's designs *vis-à-vis* Borneo were at the point of erupting into Great Britain's own undeclared war in the Far East. And so at the very time BRIAM staff members were repulsed by what they saw happening in Vietnam's high politics, Malaysia beckoned them to return to their own stomping grounds. And so in Part V we analyze the underlying rationale behind the British/GOM decision to stand down in Vietnam and redeploy the Senoi Praaq cadres onwards to Borneo.

This move essentially gets the Senoi Praaq troopers out of the "dog house" they found themselves in prior to their Vietnam deployment and it takes them one step closer to the attainment of an accepted place in the Malaysian order of battle.

Chapter 15

The Highland Environment Becomes Unstable and Unproductive

Pleiku province—in conjunction with a massive United States military buildup—rapidly developed into a virtual beehive of GVN and United States special forces activity, and from the Senoi Praaq point of view these operations invariably had an unwelcome effect on their ability to carry out that aspect of the BRIAM mission with which they had been entrusted. Such encapsulation of United States policy eventually compelled the militarization of the entire Pleiku plateau and ultimately the projection of force into neighboring Laos and Cambodia. From next to nothing in 1960 Montagnard forces grew by June 1963 to 11,000 strike force troops and 40,000 strategic hamlet militia dispersed through 40 CIDG camps and 800 Montagnard villages.[1]

DESTABILIZATION

This growth spurt represented a phenomenal increase in numbers that cannot by any means be construed in incremental terms,[2] nor as a bid at achieving comparative advantage over the communists. It was more a paradigm shift and may have looked good on paper at the time but there was no possibility—the matter of the desirability of this course of action aside—of assuring quality control.[3] The tragic results in retrospect were predictable and obviously anticipated by R.O.D. Noone given his great expertise in these matters. Hence he argued against escalation and the superfluous expenditure of material resources but to no avail and consequently the theme of guerilla against guerilla was taken to an extreme.[4] The price paid by the Montagnards for their alliance with the United States became surreal as the region was transformed into a veritable inferno of death and destruction. By the war's end—in 1975—200,000 Montagnard dead and the destruction of eighty-five percent of their ancestral villages paved a grisly road to their own small corner of hell.[5]

More immediately though once the existing primitive social order broke down, and as the geography of the Montagnards began to change, the Senoi Praaq troopers lost their effectiveness. The war in Malaya had been disruptive to their own lifestyles and the Orang Asli generally were not without experience in this type of affair. However, the situation in Vietnam was another matter altogether. The environment had become much too fluid in a sanguine tactical sense for the Senoi Praaq mission to have much import from a strategic standpoint. For that matter as the war escalated comparisons between the two situations could no longer be made with much ease.

Prior to this point the paramount appreciable difference between the two wars had been a matter of density and intensity.[6] True, other dissimilarities between the two conflicts existed but they tended to involve the usual geographic, religious, ethnic or historical considerations. Malaya was a somewhat isolated peninsula whose principal antagonists were Malays and Chinese, both subject to a well-established colonial government that was an institution in of itself. Whereas Vietnam shared ill-defined borders with Cambodia and Laos, and the war there pitted Vietnamese against Vietnamese in a conflict between the two Vietnam's the departing colonial power left in place. Thus even though some Catholics were thrown into a by and large Buddhist mix, true religious polarization did not exist let alone along strict ethnic lines as in Malaya. But these divergent conflicts between patterns were no big deal at the time. One jungle war was much the same as another if handled correctly or so people thought.[7]

Not surprisingly, the performance of the Senoi Praaq troopers was quickly impaired through no fault of their own. Their real forte lay in the relative ease with which they approached Montagnard groups and induced them to spy on infiltrators and report any Viet Cong and NVA contacts. Ideally, the GVN security program essentially provided for the buildup of an intelligence gathering apparatus to put the Pleiku plateau area under surveillance in order to ensure continued GVN control throughout the central highlands. In a perfect world it would have enabled GVN to identify and interdict the movement of DRV forces southwards along the Ho Chi Minh trail.

To accomplish the British prescribed security objective Senoi Praaq operatives first had to establish contact with the Montagnard tribes applying a pat formula to win over their leaders and then to use the Montagnards as a shield for GVN rangers —as opposed to regular ARVN forces—who in the company of Senoi Praaq and select Montagnard irregulars would seek out and close with the enemy. Naturally they did not see the recruitment of just about every able-bodied male in the highlands as a particularly desirous course of action—nor did the GVN, given the longstanding Vietnamese officials' predilection to distrust the Montagnards. To have an army of Montagnards placed at their backs held out little appeal.[8] Thus, to arm, train, equip and marshal a horde of tribal fighters ran against the grain of British experience in jungle warfare. To the Vietnamese it was akin to handing a gun to a stranger. True this sort of paramilitary unit could be very useful in the preparation of small unit ambushes but irregular forces were not meant to be employed broadly in conventional battles.

IRREGULAR FORCES MISUSED

For that matter the Montagnards tended to react adversely to some types of physical and psychological pressure that lowland peoples were more inclined to tolerate. The latter like true peasants could be put into uniform and ordered to fight, and they would do it—albeit badly—because they had never lived 'free in the forest' at liberty to think for themselves. It was a blunder for United States military officers to employ Montagnards along the lines of conscript soldiers. Once mobilized it was also an error in judgement to move the Montagnards any great distance from their own villages or tribal areas less they loose the one motivation that might be readily exploited—the instinct to protect their own. That the United States government made this mistake is freely conceded by Gilbert Layton—William Colby's man on the ground. However, he puts the onus for this and other misuse of tribal forces in the wake of Operation Switchback squarely upon the shoulders of the regular United States military.[9]

True the military tried to make over the CIDG program and turn it into an ambitious kill operation but by this point they already had a much bigger problem to tackle. The CIA's accelerated development of forces to resist the Viet Cong had contributed weightily to SVN's destabilization. Almost any group, not just those in the central highlands, willing to fight the communists was a friend in theory. The CIA, the United States Overseas Mission and the Military Assistance Advisory Group (MAAG) gave out guns to just about anyone willing to fight the Viet Cong.[10] They armed pirates, religious sects, warlords and Montagnards alike. However, their cure was more troublesome in the short run than the disease itself.

These misguided strategies early on in the conflict destabilized South Vietnam to a sufficient degree that Viet Cong infiltration was made to look like a promising tactic for the DRVN.[11] No doubt the communists had designs along these lines all along but certainly the many distress signals they picked up helped to egg them on. Thus, later day statements attributed to CIA apologists to the effect that the military turned their programs into a giant kill operation are something of an effort to shirk their own responsibility for a disaster the agency helped to create.[12]

Conversely, the Senoi Praaq formula as applied in Malaya especially late in the Emergency evolved into an array of independent actions but the scale of force in any particular engagement rarely rose above squadron strength. However, the overall concept involving a carefully calculated division of labor in which indigenous troops served as assassin *cum* reconnaissance specialists and acted as the eyes and ears of larger units was totally abused in Vietnam's highlands. Their new American friends not only turned the Montagnards into "pointers" for step-up parties they also let them be used as "strikers" through the creation of numerous mike forces. The mike forces specialized in vertical helicopter born assault and they typically operated at battalion strength of about 500 men each. These forces, designed in theory as a strategic reserve, were expected to harass the Viet Cong and NVA units infiltrating areas along the Ho Chi Minh trail. Unfortunately, however, as the conflict escalated and infiltration numbers grew the mike forces spent more and more time simply reacting to attacks against the tenuous strategic hamlets spread across the high plateau. Needless to say, Montagnard striker casualties ran high.

Nor did the Malaya counterinsurgency formula ever envision Senoi Praaq troopers functioning as social workers capable of undoing the damage inflicted by GVN venality and by the blundering of United States forces. Public relations and welfare assistance were indeed part of their portfolio but only in the sense that these services were aligned with efforts to woo Montagnards away from Viet Cong influences. On the other hand, the Montagnard troops raised by the Green Berets were able to do limited combat duty and to serve as a buffer of sorts; however, once their own tribal areas became unstable they failed either to generate useful intelligence or to repair the disruption inflicted upon their tribal way of life. The situation had simply become too fluid. Good intelligence was the keystone of the art of irregular warfare. Hence, the United States never recouped the prized information it lost first through the misuse of Senoi Praaq troopers or later through the employment of larger and larger numbers of tribal forces under its own direct control.

THE MONTAGNARDS GO SOUR

The Americans also learned the hard way in Vietnam like the British before them in Malaya that tribal peoples were a very sensitive—even fickle—lot. Ever alert to changes in their environment they were prone to switch sides at the slightest indication of weakness or incapacity. Not many outsiders were really capable of working this resource successfully and as a consequence quite a few Montagnards ran with the hare and hunted with the hounds.

By mid-October 1961 State Department documents reported most Montagnards in the Pleiku area were under the influence of the Viet Cong.[13] Pleiku area commanders—United States and Vietnamese alike—readily acknowledged most of the province was controlled physically by the Viet Cong and their Montagnard sympathizers. At best a ten to fifteen mile radius around Pleiku city was safe during the day and none of the area fit to travel in at night.[14] The Senoi Praaq troopers nevertheless had as we know some initial success working with the Montagnard tribes.

However, the artillery parks set up around Pleiku city and poor living conditions within the strategic hamlets soon caused problems for Senoi Praaq operations but GVN officers made matters much worse by their persistent abuse of those few Montagnards who were willing to inform on the Viet Cong. They habitually tortured the tribesmen who volunteered information in the hope of extracting even more. This same rationale underlay the artillery interdiction program: if a tribesman knew something about the VC or was thought to be in their presence, then he must have something to do with the enemy through guilt by association.[15] Such tactics were akin to killing the golden goose for bad faith while she was still laying eggs. Unfortunately, it was not so easy to be venal and successful at the same time and the gradual deterioration of events through August to November 1963 all but sealed the Malayan team's fate.

Nonetheless, Senoi Praaq assessment of the danger the Viet Cong posed even at the twilight of their mission, particularly in the face of such obvious GVN abuses, were at odds with the appraisals supplied by United States and Vietnamese officials. Through sheer determination the Senoi Praaq troopers were able to maintain a high degree of mobility for some time despite eventual pressure to go to ground and become tied to fixed positions. Unfortunately, their momentum eventually stalled in light of the existent circumstances and they lost freedom of movement.

Bureaucracy and unfriendly officialdom made the play on personalities vital and necessitated super human efforts to get results in setting up logistic supplies for Senoi Praaq needs. This ordeal became a paper war within a shooting war. Soon Vietnamese interference made the secret training programs all but impossible to carry out and it became extremely dangerous for the Senoi Praaq to circulate in their designated areas. Too bad the people involved—South Vietnamese, Montagnard, British, Malay and American—didn't have more use for one another.

No doubt given time and left to their own devices the Senoi Praaq troopers undoubtably could have cast a much wider net in the central highlands without losing so much of the catch. They had better local intelligence than the United States and RVN operatives and they also had a much higher retention rate among tribesmen won over to the anti-communist cause.

TROUBLE IN TRIOS

In face of the steadily rising tempo of the insurgency throughout South Vietnam Robert Thompson was unwavering in his firm belief the RVN could overcome the communist assaults if the Montagnards stood firm in the central highlands. His haunting words "... if the Montagnards did not go sour on us ..." were repeated to Ambassador Lodge on September 12, 1963, and then dispatched to London a short while later.[16] Sir Robert's epigrammatic message signaled the portentous defeat that lay ahead.[17] The Montagnards did go sour, opening the door to Saigon by way of the Ho Chi Minh trail, the DRVN's main route for military supply which snaked along the Laos and Cambodia frontiers into South Vietnam's high plateau.

There is a dearth of information about this drastic turn of events in the highlands in the early 1960s in general and the resulting abortion of the secretive Senoi Praaq mission in particular. However, the United States and British diplomatic correspondence of the period pinpoints three separate but interrelated actions which help explain the rapid ascension of the Viet Cong throughout the plateau area: RVN's inveterate mismanagement of the Montagnards; the Senoi Praaq's unheralded, yet eye-opening withdrawal; and the United States military's premature draw-off of special forces contingents for redeployment closer to Laos and Cambodia which left the VC free to operate in their rear.[18] Similarly, some DRVN forces in the Mekong Delta region south of Saigon, badly mauled by aggressive ARVN attacks, beat a retreat to the highlands at about the same time.[19]

The wisdom of the United States strategy aside it spawned the very conditions which created VC/NVA counter forces at the same time it afforded them a vacuum in which to operate. Thus, the high plateau if not all Vietnam was effectively lost well before RVN's ultimate collapse in 1975.

Chapter 16

Impetus for the Senoi Praaq Withdrawal

Death itself is a process and the pathology of the Vietnam War is well known to have been a morbid one. Any number of unsavory causes may be bandied about as catalysts for the sickening rout of United States and RVN armed forces. However, the specifics of RVN's passing as they relate to the Senoi Praaq experience and the abortion of BRIAM remain to be established. In this respect we can point to three critical events which hastened the British decision to stand-down in South Vietnam: The debacle at Plei M'rong, President Ngo Dinh Diem's assassination and the outbreak of hostilities between Malaya and Indonesia on the Island of Borneo.

PLEI M'RONG RINGS IN THE NEW YEAR

Both the quartering of the Montagnards in the strategic hamlets program and the CIDG levy were indeed dubious gambits but they did provoke a vicious DRV response. Whether or not the fiery counteraction by the communist forces may be cited as belated testimony to the effectiveness of such programming in the coherent prosecution of the war is an open-ended question. In any event the Viet Cong and NVA employed brute force tactics in countless attempts to infiltrate and compromise both the strategic hamlets and the CIDG.

Communist forces with the help of turncoat highlanders briefly overran the Montagnard training center at Plei M'rong in January 1963. Plei M'rong, near Kontum, was an important fortified military camp, not the usual strategic hamlet protected by poorly trained militia. Nonetheless the raiders killed and plundered at will, snatched the hapless for their usual fun at a more leisurely moment and then quickly made good their escape. It would be a real understatement to say the place was a mess when they left.

A section of the camp, to the defenders credit, did manage at a cost of great human sacrifice to hold out and order was soon reestablished. Yet, that a large VC/NVA unit could sneak up close enough to mount an audacious assault so near

Pleiku city was a very bad sign. Indeed, the whole point of using indigenous elements in irregular forces is to have one's finger on the pulse in any given geographic area. Montagnard duplicity in the matter spelt disaster. Their native instincts foretold the oncoming communist victory.

And so by the time Noone's troopers actually deployed in the highlands in early 1963 some Montagnards had already begun to reposition themselves in anticipation of an eventual GVN and United States defeat. This was clear evidence of their classic self-defense mechanism at work and an example of survival strategy in action. They would not be trapped in a collective sense between the Scylla of Viet Cong intransigence and the Charybdis of United States escalation to the exclusion of one danger over another. However, for the individual plateau tribesman transition from passive to active Viet Cong supporter signaled all too often his doom and became an ill-fated first step down the road to oblivion.

It was the end of the beginning. The beginning of the end would come a little more than two years later with another sharp VC/NVA assault upon the airfield at Pleiku City proper and the nearby United States military advisor's compound at Camp Holloway. On February 7, 1965 at 2:38 p.m. nine Americans were killed and seventy-six others wounded in a bold, well coordinated attack. Additional hits were taken elsewhere in the II Corps region and again at Qui Nhon on February 10th. These episodes together triggered United States ground escalation by conventional forces[20] and the air war over North Vietnam a short time later.[21]

British consular officials in Saigon recognized immediately the significance of the Plei M'rong assault and suggested " . . . this attack could be the pattern for others to come."[22] Their suspicions about the deteriorating Montagnard situation were duly aroused. They would not have long to wait for further evidence to start to accumulate in the form of a string of incidents leading up to the now infamous Pleiku/Qui Nhon attacks. The Viet Cong made no less than ninety-four probes and eleven attempted infiltrations against Special Forces camps in the first six months after Plei M'rong. CIDG patrols reportedly encountered Viet Cong formations on more than 300 occasions (i.e., 374 contacts) in the same period.[23] During this brief period the tone of the mission's cable traffic (conducted through the British Embassy) with the Foreign Office in London started to change. Message content began to adopt a slow discernable drift away from the positive "how to win the war" tone to the matter of finding a convenient means whereby French leave might be taken on the Americans.

Senoi Praaq veterans present at ground zero itself for several similar incidents have related to the author they already knew things had begun to go badly. Up to this point they reasoned the Viet Cong had simply limited the effectiveness of their mission somewhat but they did not consider themselves to be put in immediate danger by the turn of events. However, in the light of these most recent developments they were no longer safe to live in tent cities while not out on operations and instead they had to spend too much time underground in bunkers at their base camps. This was a novel situation for them as in Malaya, the Senoi Praaq were the ones who made the communists live underground and so their roles were reversed.

Needless to say this juxtaposition of roles was not natural and to their way of thinking it came about because there was no longer that much space between the bullets flashing around Pleiku. The Viet Cong and NVA had their number.

But R.O.D. Noone himself was no quitter. Given his passion for CRW it would take more than one incident to induce any acknowledgment of defeat on his part. Unfortunately, the straw that broke the camel's back would not be long in coming. President Diem along with his brother Nhu was soon to be deposed and murdered in a United States supported *coup d'etat*. As a consequence the problems the Senoi Praaq had encountered heretofore with the Vietnamese authorities concerning the Montagnard program were about to be exacerbated.

NGO DINH DIEM IS ASSASSINATED

Ngo Dinh Diem's assassination in November 1963 provided a welcome window through which to exit the festering quagmire in South Vietnam. Soon after President Diem's capture and secretive execution the brain behind Senoi Praaq logistics—Ahmad bin Khamis—and a second and larger party of fighters had their marching orders to Vietnam abruptly cancelled.[24] This directive precluded any form of meaningful BRIAM escalation on the ground as an effective Senoi Praaq buildup could not take place in such an unstable political environment.

Both the United States Department of State and the American press had once heaped lavish praise on Diem, the stoic, reclusive Catholic layman and former officer in the French colonial service, when he was named premier in the government of the new RVN in 1954. Fresh from his sojourn in the United States at a seminary in New Jersey and brief employment at Michigan State University to assist in the preparation of a major aid project focusing on police training and public administration education, Diem won plaudits for his moral courage and strong anti-communist stance. However, as President he had become very unpopular in the American media and his regime drew a lot of verbal flack that translated into negative political pressures on policy makers back in Washington. Incidents of self-immolation by Buddhist monks, arbitrary arrest, and torture of political prisoners were an unpleasant business. The mouthy Madame Ngo Dinh Nhu and her gleeful talk of barbeque did not make for a good impression either and the GVN would have been better served by a more skillful apologist.

United States representatives on the scene in Saigon were also very unhappy about their own inability to get the Diem government to conform to their view about how to develop the economy and prosecute the war.[25] Diem had his own parochial way of doing things and the arrangements he put in place were quite feudal in character. Moreover, corruption and nepotism ran rampant throughout the country. Deal makers routinely payed twice for the same service—once through legitimate channels to the GVN and then a second payment under the table to officials operating at the ministerial or provincial levels.

Money would just not buy some things. For instance, GVN's full support for expansion of the Montagnard program could not be had at any price no matter how hard United States officials tried.[26] As we know the Vietnamese did not trust the Montagnards. Furthermore, the BRIAM prescription provided for small-scale, low technology assistance with minimum intrusion. The Americans were also pressing President Diem at this time for an increase in the ceiling of United States forces in Vietnam.[27] However, he made every effort to keep a lid on the expansion of MAAG and see that MAAG did not evolve into the Military Assistance Command, Vieetnam (MACV). Diem undoubtedly was influenced by Robert Thompson and Malaysia's Tunku on this point. They advocated large infusions of economic aid and a strengthened civilian police operating from behind a screen of military support.

To make matters worse Diem was also rumored to be a voyeur and a pedofile who kept lewd displays of pornography strewn about coffee tables in his personal quarters at the presidential palace.[28] His director of an extensive secret police apparatus, Dr. Tran Kim Tuyen, maintained lurid dossiers of sexually compromising pictures and materials to keep Diem's lieutenants in line.[29] Such droll tastes contrasted sharply with the outward puritanical machinations of a regime in which dancing and popular Vietnamese love songs were banned. Diem was an odd duck whose tenure was paradoxical.

No doubt United States intelligence, diplomatic and military people in Saigon were besides themselves at being jerked around by a political leader some operatives described as a "flaky little twerp" whose high perch rested upon connections via the Vatican to Washington. The United States had spent more than $500,000,000 yearly for nearly a decade in support of a regime that had become an international embarrassment.[30] Nonetheless, Robert Thompson appealed directly to President Kennedy to stay the course with Diem. However, Ambassador Lodge and his associates deemed the situation to be unbearable and they wanted it resolved post-haste.[31]

Consequently, spin doctors in the loop acknowledged the implausibility of an inflexible Catholic regime ever functioning effectively in an essentially Buddhist culture and shamelessly began to make a lot of noise of their own around Saigon and in Washington about the absence of democratic institutions in Vietnam and Diem's lack of popular support among the Vietnamese people. It was the same old tune they sang whenever it was time to get rid of someone who would not toe their policy.[32] Diem was an embarrassment who just had to go and so one fine day early in November 1963 not long after surrendering to the Vietnamese military subsequent to a second coup attempt he and his brother Nhu were killed on the sly reportedly by certain parties with linkages to United States operatives in Saigon.[33]

THE HIGHLAND SITUATION COMES UNGLUED, BORNEO BECKONS THE SENOI PRAAQ

The political environment in Vietnam was a very unstable one in which corruption, connections and patronage were the order of the day. British officials on the scene kept trying to impress upon the Americans the war could not be won in the short term and they reasoned that in the long term Diem's regime did not really matter.[34] They preferred to make do with Diem for a time longer and to let him whither on the vine rather than cut him off at the knees. Some came to wonder among themselves if the Americans had any idea at all about what they were doing in Vietnam.[35]

Diem was the one politician the British thought capable of holding South Vietnam together despite his many personal deficiencies and professional shortcomings not to mention his woeful lack of popular support. Consequently they preferred to play it safe and work within the rules Diem had set rather than risk rule by the young ARVN generals. We now know they were correct in their assumptions. In short order cronyism and corruption were surpassed by confusion and real incompetence.

A flurry of coups followed Diem's assassination as a clutch of South Vietnamese generals jockeyed for position in the resultant political vacuum.[36] The generals neither formulated a coherent war policy nor did anyone of them keep power long enough to provide the necessary continuity for its management. The resulting inertia and confusion completely stymied the BRIAM initiative and the military situation deteriorated steadily thereafter.[37]

The negative impact of Diem's removal upon BRIAM's Montagnard program in the highlands was immediate and debilitative to the extreme. This should come as no surprise as we do indeed know the Vietnamese officials on the scene had always been somewhat obstructionist to say the least due to their own problems with the Montagnards. To offset the distance they kept from the Montagnards they were in the habit of fobbing off interpreters and liaison officers upon the Senoi Praaq in an effort to keep abreast of developments. But now, in light of the assassination of Diem and Nhu and in the wake of a surreptitious buildup of tribal forces by United States advisors and special forces personnel the Vietnamese were apoplectic. Indeed sometime in early 1963 Nhu himself during a visit to Pleiku was reported to have had quite a shock upon receipt of the news about the extent of the buildup.[38]

The Diem government officials had become "shit scared" the Americans would really get to know the Montagnards and gain their loyalty. This was a legitimate fear on their part given the extent of previous United States overtures along these very lines. The United States had already expanded the Montagnard forces behind GVN's back despite the bad feelings this would create. The British on the other hand had been only too happy to press President Diem on other matters and thus to pass off as impractical any massive Montagnard buildup.[39] The British unlike the Americans were indeed fairly sensitive to GVN concerns on this score and inclined to tread softly.[40] The Malayan experience and similar but less stringent Malay concerns about arming the Orang Asli were a recent memory and both R.O.D. Noone and Sir Robert could be very tactful.

Alas, by the time of Diem's demise it was too late to do any more good in South Vietnam. The breakdown of law and order subsequent to the assassination was the straw that broke the camels back and precipitated the withdrawal of R.O.D.

Noone's Malayan team. British officials didn't need a graph or a regression line pointing to imminent disaster to know beyond a shadow of a doubt all was lost in the highlands. The Montagnard situation around Pleiku had turned into the type scenario bad dreams were made of and threatened to consume the Senoi Praaq. No point in hanging around to be drained dry by Minerva's *succubi.*

In sum Malaya and Great Britain had done what they could for South Vietnam and the United States and they wisely chose to decamp in the wake of Diem's assassination. However, the dirty deed was not sufficient justification for them to abruptly call it quits, cut their losses and leave the Americans in the lurch. BRIAM needed another reason to ditch the highland program and Malaya's confrontation with Indonesia presented an excellent opportunity around which to model her exit strategy and extract the Senoi Praaq.

Chapter 17

When the Going Gets Tough, the Tough Get Going

For the British discretion truly must have been the better part of valor. Their internal correspondence setting up BRIAM's dissolution was so smooth one wonders if these documents were the products of the bureaucracy or works of art. Analysis of their content reveals a program of deception in which the line between complex productive statements and dubious syllogisms is blurred to say the least. The shame of it all though was that the success of the project might otherwise have changed the course of events which eventually led to such an obscene level of direct US involvement in VN.

BAIT AND SWITCH

If the best way to lie to someone is to tell half the truth British officials certainly were most generous with their American friends when they gave them the short end of the stick. Her Majesty's representatives to Vietnam managed through organization theory type arguments to convince their American colleagues that it was in their own best interests to merge BRIAM's advisory function into the United States Operations Mission (USOM).[41] Towards this end they touted enhanced efficiency and unity of purpose as a panacea for Vietnam's problems and as the means whereby Great Britain's contribution might be optimized.

These smoothies no doubt realized only too well the best way to make an operation disappear was to ensconce it within a larger organization. Human nature could then be counted upon to take care of the rest as people love to muscle in on each others action.[42] Thus, through consolidation they pulled off a bait and switch manoeuver in which BRIAM and the Senoi Praaq were withdrawn and replaced by a handful of office wallahs.[43] True these new recruits were ex-colonial police officers with Malayan Emergency experience but they were not by any stretch of the imagination a substitute for a cadre of hardcore indigenous fighters. These lesser commitments themselves could then be withdrawn piecemeal without any fuss a few years later once the contracts of the individuals involved expired.

The suspect claim of the need for restructure and consolidation of advisory assistance in Vietnam was amply offset by the fighting on Borneo that was about to spread to peninsular Malaya proper. British and Malayan forces had suffered several setbacks that required immediate attention by R.O.D. Noone, et al. The British clearly had a pressing need to redirect their energies away from Vietnam. So, they were able to effect these changes masking their own loss of enthusiasm for the Vietnam War and at the same time they kept intact American approbation of their policy of resistance to Indonesia. In the final analysis what the British officials told the Americans really did ring true. Their motivations had an air of authenticity about them but it is hard to let pass without comment on some telling words in their internal communiques: " . . . Our real motive must be . . . we cannot afford to let the Americans believe . . . "[44]

GRASPING AT STRAWS

However, the greatest irony of Great Britain's experience in Vietnam is that only sometime later after the withdrawal of the Senoi Praaq troopers did the United States try to make any real use of what BRIAM best had to offer all along—R.G.K. Thompson's genius for counterinsurgency, irregular warfare specialists, techniques for pacification, and a model for amicable relations with the indigenous peoples across the high plateau region. Instead the United States military officers largely ignored BRIAM advisory personnel in Saigon for nearly three years ostensibly because they had had a negative experience with third country elements in the recent war in Korea. Numerous United Nations member countries sent token forces to this war to fight but the prevailing wisdom among United States military commanders was they required too much "care and feeding" to be worthwhile.[45]

Obviously, this general assumption on the part of conventional military authorities was erroneous in the case of the Senoi Praaq. More substantive exposure to and a better grasp of the particulars of the Malayan Emergency would have provided them the information they needed about the prerequisites for effective jungle warfare. Only continued failures subsequent to the introduction of their own ground forces in great numbers finally pushed the American military establishment to turn in desperation to an asset that had been theirs all along. Thus, it was late in the day that the question about whether or not Sir Robert should be brought onboard as an informal member of the American team came up in high government circles.[46]

Too bad there had not been that much direct cooperation between the two advisory missions while BRIAM was in full swing. Particularly so given that similar friction did not exist at the ground level between the Senoi Praaq per se and Americans operating in the field.[47] BRIAM/United States interaction at the policy-making level tended, as a practical matter, not to exist. Instead ranking Americans harbored a good deal of rancor over the fact Robert Thompson was accredited directly to President Diem and his prescriptions often contradicted with their own.[48]

The British for their part had no intention of usurping GVN authority. They tried to flatter and cajole the Diem government into action whereas United States operatives sought to direct the war themselves. Thus, BRIAM and MAAG officials dealt separately with the GVN and continued to do so even after Diem's assassination. Hence, a fair amount of their energy *vis-à-vis* interaction between allies was largely confined to amicable chats among VIPs. These same individuals then would cast dispersions upon one another in private reports to their peers.

Ambassador Nolting in his reports back to Washington was quick to pounce upon Sir Robert and Ambassador Hohler for their going straight to Diem. Despite the existence of friction Sir Robert is said to have responded always with good grace to the American team position, although he was clearly annoyed with Ambassador Nolting.[49] Discussions late in the day between Sir Robert and Ambassador Henry Cabot Lodge are also a fitting example of this tendency—particularly in regards to the fate of President Diem. These two men would never dirty their own hands as the situation was much too stratified but it is only too clear they did meet to discuss the merits of Diem's disposal.[50]

BRIAM FADES AWAY

Vestiges of BRIAM were thus destined to remain in the country during the United States escalation but in 1965 Robert Thompson left Saigon a disappointed man. He would make yearly trips back to Vietnam in the future and do quite a bit of analysis for Presidents Johnson and Nixon but to no avail. Despite an outward show of optimism on Sir Robert's part Humpty Dumpty could not be put back together again at any cost. For all intents and purposes active BRIAM/Senoi Praaq participation in the Vietnam war ended at this time.[51]

However, Senoi Praaq operatives continued to have some institutionalized interaction with the Vietnamese and presumably a few Montagnards through their instructional work at the JWS in Kota Tingi, Johore.[52] The British and Malays did not make their Vietnamese friends go cold turkey on this aspect of BRIAM's mission. Instead BRIAM's training function was shipped offshore to Malaya where it was sanitized in the sense that such activities no longer involved actual combat.[53]

Senoi Praaq personnel assigned to the school on a rotational basis, ran JTO segments for the many soldiers who sought a coveted certification in long-range tracking. The JWS did big business back in those days and catered to a variety of armed forces—British, French, American, Australian and Vietnamese are but to name a few. However, the ARVN through the United States military assistance program supplied a majority of the customers enrolled in offshore training at this facility and thus the Vietnamese military did not experience immediate withdrawal symptoms upon BRIAM's departure from Vietnam.[54]

THE SENOI PRAAQ ENLIST IN A NEW WAR AND REDEPLOY TO BORNEO

Hence, it was in a poisoned overall atmosphere of ill will, mistrust and war that the Senoi Praaq said goodby to their many new Montagnard friends and set out for

Borneo. They were to a man glad to move on and their only regrets were those shown in the faces of the tribesmen they left behind. However, any sense of relief the men may have felt at being withdrawn from harms way would quickly evaporate. They were about to jump from one frying pan into another. Great Britain had become embroiled in yet another of her own undeclared wars and Senoi Praaq cadres would shortly find themselves in action on yet another exotic battlefield.

The Vietnam conflict albeit very well-known was not the only messy little war heating up in Southeast Asia at that particular moment. Nor was it a stand alone event isolated in time and space from other confrontations in the region. All these engagements were interrelated in the sense that they were very much a part of the cold war and as such harbingers of a sea change in the international world order. Thus each was important in its own respect although some were less famous than others.[55]

Britain's clash with Indonesia was an excellent example of how the specter of Vietnam could completely overshadow a lessor conflict both in the world press and the history books. Borneo and peninsular Malaysia were on the fringes of the United States sphere of influence in the Pacific area and there was a tendency to ignore the fact that another world existed just to the south of Vietnam. However, to the British the rich territories of Sabah and Sarawak represented the epiphany of afterglow shed by an empire upon which the sun had already set.

Britain would thus seek to preserve regional stability so that existent linkages between Malaya and the mother country might be kept intact. Consequently, substantial territorial conglomeration took place throughout her Far East dominions and post-independence era Malaya in particular experienced such clustering. The fledgling state attempted through Foreign Office prompting to join with Singapore, Sabah and Sarawak to form the Federation of Malaysia.

However, this maneuver provoked serious political and military repercussions and it did not go unopposed by other countries in the region. Malaya's neighbor Indonesia, in particular, felt threatened by the prospect of a greater Malaysia and was leery of Britain's new Commonwealth of Nations. Unlike in Malaya, in which the Malay/British relationship had been good, the Indonesians had a grating experience with the Dutch.[56] Indonesia quickly accused Britain of neo-colonialism and began to explore the most appropriate physical means by which to resist formation of the Federation.

To complicate matters further Indonesia's own proclivity towards expansionism dictated a conflict then ensue. Thus, once British and Malay determination in the matter became manifestly clear the Sukarno government, necessarily to the Indonesian way of thinking, responded by force of arms to the formation of the new Federation of Malaysia. At stake on Borneo island were the British Crown Colonies of Sabah and Sarawak and the Sultanate of Brunei. The seizure of these territories would give Indonesia control over the lesser part of the island beyond Kalimantan which she did not already own. For that matter these territories possessed vast reservoirs of oil, abundant mineral resources and some of the best tropical timber to be found in Asia.

Hence, in late 1962 the Sukarno government provoked the puppet North Kalimantan National Army (NKNA) to make a bid at seizure of Brunei and parts of Sarawak. GOM and British forces easily crushed the NKNA rebels by early 1963 but these actions escalated later the same year into fierce clashes between regular Indonesian forces and Commonwealth troops. The lengthy Sabah/Sarawak border with Indonesia became a veritable war zone across which many Indonesian raids were launched in force. There was every possibility that Indonesian troops might overrun the remainder of the island and throw the British and their allies into the sea or otherwise drive them away just as they had done the Dutch.[57]

Part V: Notes

1. Kelly, 37. That 1962-1963 represented an early turning point or "a race between buildups" is further indicated by John Prados. Se Prados, *Blood Road*, 56-57. Gilbert Layton—roughly the American equivalent in Vietnam of R. O. D. Noone in Malaya—lumps these forces together and places this figure at an aggregate sum of about 36,000 men most of whom were tribal peoples. Most were recruited withing the space of less than a year. See Sedgwick Tourison, *Secret Army Secret War: Washington's Tragic Spy Operation in North Vietnam* (Annapolis: Naval Institute Press, 1995), 24.

2. It is also hard to imagine that a modest initial American presence could expand to such ungodly proportions. The Green Berets sent to South Vietnam (Special Forces detachment A-35) arrived in late1961. They were divided into two groups of approximately seven or so men each (whether or not they arrived as one or two groups is unclear). However, one-half went to Buon Eano and the other to the Hoa Cam training center in Da Nang. Both groups were in place by December of that same year. See Kelly, 25, 32-33.

3. The Montagnard buildup is also reported to have brought about the institutional decline of the Green Berets as a quality special forces operation. The unit expanded much too quickly to meet requisite manpower needs and was sucked into the same damming vicious cycle. Entrance requirements changed—cross training, language requirements—were dropped and too many cowboys were allowed into the program. See Thomas Bailey and Bruce Lawlor, "The Phoenix" in *Everything We Had: An Oral History of the Vietnam War by Thirty-three American Soldiers Who Fought It*, ed. Albert Santoli (New York: Random House, 1981), 199-202.

4. Noone's prescriptions made their way into Sir Robert's own reports on BRIAM progress and ultimately he also came out against the hasty escalation of counterinsurgency operations. Sir Robert felt far too much fuss had been made over the guerilla versus guerilla concept. He even warned President Diem in a conversation that if "... the main emphasis is placed merely on killing terrorists there is a grave risk that more Communists will be created than are killed." See Anthony Short, *The Origins of the Vietnam War*. New York: Longman Inc., 256. In this excerpt Mr. Short only quotes Sir Robert briefly on this meeting between he and President Diem. For the rest of the discussion, as referred to in Short's footnote, see Thompson to Diem, 13 November 1961, *US-Vietnam Relations*, Book 11, p. 347 (*Foreign Relations of the United States*).

5. These figures firmly place the Montagnard tragedy in the category of the blood bath

that took place in Cambodia upon the Khmer Rouge ascent to power. The great shame at the betrayal of the Montagnard peoples by the United States Department of State is also undeniable. But unfortunately time goes on and the *post facto* quip "What is a Montagnard? Something you eat or drive?" by a glib State Department employee some years later points up the worth of their sacrifice. It thus bares repeating, albeit sadly, that only a handful of tribesmen ever made it to the United States. See Josh Getlin, "Celebrates First Thanksgiving; Montagnard Refugees Find New Life in U.S.," *Los Angles Times*, 30 November 1986, 1.

6. The Vietnamese people were thick on the ground in Vietnam and, for that matter, unlike in Malaya— where the British never had to deal with hostile masses of humanity and did not fight a monstrous Dien Bien Phu type battle which would so obviously cast the dye —traumatized for all intents and purposes by the French.

7. Had Sir Robert's prescriptions not been misapplied in form, function and scope there is every possibility that the Vietnam War could well have taken a different course. To this day protagonists in the British camp insist the outcome would most likely have been otherwise had the war been kept small and never escalated to the extreme.

8. "The Americans have put an army at my back" is General Ton That Dinh's turn of phrase as he told Ngo Dinh Nhu of the extent of the American build up among the Montagnards at a tribal celebration held in Buon Enao. 18,000 Montagnards had been armed and this fact had been "a deep secret" according to Lucien Conein. See Hickey, 80.

9. Tourison, 74.

10. Memorandum from Michael V. Forrestal to Seymour J. Janow at USAID, February 8, 1963. Hilsman Collection, Box #3, File: Countries-Vietnam [2/1/63-8/21/63], 02/8/63, John F. Kennedy Library.

11. It is only too popular in the literature to harp about the CIA's incompetence and ineptitude. For a recent example see Richard H. Shultz, *The Secret War Against Hanoi: Kennedy's and Johnson's Use of Spies, Saboteurs and Covert Warriors in North Vietnam* (New York: Harper Collins Publishers Inc., 1999), 29. This criticism is not new nor is it always so civil and analytical. For a blunt and more derogatory description in which CIA personnel are labeled "incompetent drunks" see Mike Beamon, "The Green-Faced Frogmen," in *Everything We Had: An Oral History of the Vietnam War by Thirty-three American Soldiers Who Fought It*, ed. Albert Santoli (New York: Random House, 1981), 213. To be fair though one should differentiate between career officers who could do much better and those men and women in the field who do most of the leg work in confrontation type situations like Vietnam. In such cases the individuals involved at the ground level are often contract employees, detached from policy making and pawns in their own right. On the other hand some career officers in the CIA, USAID and at State, like toadies, simply subsist off a line in the budget and spend their time rotating back and forth between Washington, D.C. and various embassies abroad where they circulate mostly among a closed circle of people similarly engaged at foreign postings. They often do their real work through intermediaries—unsuspecting people, often American citizens bereft of diplomatic immunity, who are cultivated for the purposes of fronting for them—but these supine techniques all too frequently fail to get the job done.

12. The CIA sponsored programs which were amalgamated into the Civilian Irregular Defense Group (CIDG) program by way of the Village Defense Program were not strictly "political" as opposed to "military" activities. They entertained a fair amount of killing in their own right albeit on a much smaller scale. Thus, the term "dysfunctional kill operations" might better describe the new arrangement. For a good example of interagency sniping and spin control, in which antagonists air their grievances against each other through

manipulation of an academic medium see Hickey, 80.

13. Similarly, by the CIA's own estimates about ten percent of the tribal forces they recruited were Viet Cong. To their awkward logic this meant—as long as substantial Viet Cong or NVA units were not encountered in the field—that they had the enemy outnumbered by about nine to one. See Tourison, 54,55.

14. In the brief interval between January 1960, when State Department reporters were relatively free to travel in Pleiku and Khontum provinces, and October 1961 the situation had deteriorated significantly. Vice Consul John J. Helble makes this admission in a Foreign Service Despatch from Hue to Washington, concerning security conditions and tribal problems in Pleiku and Khontum provinces, dated October 12, 1961. See, Department of State, Central Files, 751K.00/.10-1261, National Archives, Washington, D.C.

15. GVN commanders (i.e., the Chief of Staff of II Corps Region) acknowledged by this time that they had alienated the Montagnards—"We have lost the Montagnards. Now we must start all over again and try to win them back." This quote is taken from the aforementioned report on security conditions and tribal problems in Pleiku and Kontum provinces by John J. Helble in Hue, dated October 12, 1961. See Department of State, Central Files, 751K.00/.10-1261, National Archives, Washington, D.C.

16. This message is conveyed by Robert Thompson to F. A Warner through the aforementioned cover letter and the attached summary of a September 12, 1963 meeting between Robert Thompson and Ambassador Lodge. See Thompson to Warner, 18 Sept. 1963, PRO, BRIAM (i.e., FO) Vietnam/371/170102. It is essentially repeated in another letter penned a short while later to E. H. Peck at the Foreign Office while Thompson was on a trip to New Zealand. In this letter though, Thompson adds "I have no very clear evidence but I strongly suspect that all is not well with the Montagnards situation." See Thompson to Peck, 9 Oct. 1963, PRO, FO New Zealand/371/170102.

17. In his summary memorandum for March 1963 Robert Thompson emphasizes manipulation of the Montagnards as a means by which to curtail RVN infiltration. He states ". . . if the Montagnard policy continues to be successful and the absorptive capacity of the Annamite Chain for such infiltration continues to be reduced, then it is unlikely to be of decisive significance." See Robert Thompson, "The Situation in South Vietnam," Mar. 11, 1963, PRO, BRIAM (i.e. FO) Vietnam/371/170100.

18. Discussion of the draw-off appears in a memorandum from Michael V. Forrestal (of the National Security Council Staff) to the Secretary of Defense. See, United States Department of State, *Foreign Relations of the United States, 1964-1968*, Volume 1, Vietnam 1964 (Washington, D.C.: Government Printing Office, 1992), 78.

19. Cable traffic from Saigon paid particular attention to enemy troop dispositions. This information appears in the aforementioned outgoing aerogram of Sept. 2, 1961 from Saigon authorized and signed by Sterling J. Cottrell. See, Department of State, Central Files, 751K.00/9-561, National Archives, Washington, D.C.

20. In sum, having failed in the guerilla versus guerilla approach United States policy makers then vacillated between extremes and allowed themselves in turn to be "... influenced by the very worst interpretation of Clausewitz and the Prussian example of Moltke ..." They became obsessed with the destruction of main force units in conventional type battles. Here Anthony Short quotes Sir Robert's writing as if to say the US had embraced the second of twin evils. See Short, *The Origins*, 256.

21. Neil Sheehan, et al., *The Pentagon Papers: as Published by the New York Times* (New York: Quadrangel Books, 1971), 352. Schultz cites nine dead and 128 wounded in *The Secret War Against Hanoi*, on page 213. Similar information— eight dead and 104 wounded—is conveyed by John Prados. See Prados, *Blood Road*, 94. Prados also observes

in a footnote on the same page that the DRVN marks February 7 as a pivotal moment in the struggle for reunification by annually celebrating this occasion (i.e., the attack on Pleiku). Thus, in so much as the United States had chosen Pleiku province as the site from which to initiate its counterinsurgency program the communists can glean a certain amount of poetic justice from the commemoration of the event.

22. This observation was made in an aforementioned report by none other than H.A.F. Hohler (Great Britain's Ambassador to the Republic of Vietnam) to Lord Home (Great Britain's Foreign Secretary). See Hohler to Home, 30 Jan. 1963, PRO, FO Vietnam/371/170100.

23. Prados, *Hidden History*, 78; *Blood Road*, 58.

24. This group would have supplemented a small party of men (inclusive of the "bad" Yusof Khan) Norman Herbolt took up to Pleiku, subsequent to his own suspension as Senoi Praaq Commander in Malyasia, in April, 1963 and it is perhaps more aptly termed a third as opposed to a second wave. In this respect, Great Britain—more or less—committed in mid to late 1962 to increase her participation in the Vietnam war. Hence, as of January 1963, the Americans expected an increase in paramilitary aid from their British allies. This revelation appears in a report to the Special Group for Counterinsurgency *vis-à-vis* Third Country Aid to South Vietnam. See United States Department of State, *Foreign Relations of the United States, 1961-1963*, Volume 3, Vietnam (Washington, D.C.: Government Printing Office,1991). 24.

25. Ambassador Lodge encouraged the coup against Diem. See Stanley Karnow, *Vietnam: A History* (New York: The Viking Press, 1983), 278. Here a rift occurred over Diem's fate between the wishes of civilian and military leaders. General Harkins, the ranking United States military officer in Vietnam is said in General Westmoreland's memoirs to have rejected adamantly any consideration of a head of state's assassination. As guests in a foreign country his military sense of honor forbid such an ill-mannered manifestation of political expediency. Such differences of opinion along ethical lines served to provoke serious tensions within the United States politico military complex. After the coup General Harkins and Ambassador Lodge were hardly speaking. General Westmoreland points a second finger at Ambassador Lodge in that Lodge is said to have rejected ideas floated by Robert Thompson with which Westmoreland happened to agree. See General William C. Westmoreland, *A Soldier Reports* (New York: Doubleday & Company, Inc., 1976), 65, 66, 69.

26. President Diem signed away much of his authority in the central highlands on May 5, 1961 but the Americans were never satisfied. See, Department of State, Central Files, 751K.00/5-961, National Archives, Washington, D.C.; six months later they continued to beat this dead horse in their internal communiques. United States Department of State, *Foreign Relations of the United States*, 1961-1963, Volume 1, Vietnam 1961, (Washington, D.C.: Government Printing Office, 1988), 699; Frederick Nolting, *From Trust to Tragedy: The Political Memoirs of Frederick Nolting, Kennedy's Ambassador to Diem's Vietnam* (New York: Praeger, 1988), 53.

27. Diem wanted a cap on the number of advisors in South Vietnam but some of his own people were swayed by pressure from Washington. Defense Minister Thuan gave General Maxwell Taylor a letter acceding to this demand. See Tourison, 78.

28. Diem's peccadillos are not thought to have been reported in the American press and do not appear in conventional accounts of the war. More innocent and politically correct descriptions of his reading habits—"Piles of American adventure magazines littered Diem's bedroom;"—are described instead. See Karnow, 310.

29. One daring top official so compromised admired the libelous photographs and

requested copies to provide his friends and associates.

30. Towards conclusion of the war the aggregate total of these expenditures would rise to an awesome 150 billion dollars. See Tourison, 254.

31. There is also a great deal of speculation in the published literature about overtures by the Nhu brothers to the communists. For examples see Hickey, 89 and Karnow, 291-292. However, these concerns are not reflected in available archival materials dated to the time and may have surfaced only as smoke to muddy the waters.

32. Once control over counterinsurgency operations had been secured the Americans then began to press President Diem on other military matters. They ostensibly wanted Diem to decentralize further control over prosecution of the war in the name of efficiency and argued that the war could not be run effectively out of the Presidential Palace. However, this rationale sounds a little self-serving when one considers that Presidents Kennedy, Johnson and Nixon are all known to have more or less made military policy in Washington. President Johnson's Tuesday lunch club, in which he and his cronies would sit around and pick targets, is a real case in point. In fact, Donovan Webster Reports President Johnson was known to boast "they can't even bomb an outhouse without my approval." (See Donovan Webster, *Aftermath: The Remnants of War* (New York: Pantheon Books, 1996), 174.) For Ambassador Nolting's take on the situation see, United States Department of State, *Foreign Relations of the United States, 1961-1963*, Volume 1, Vietnam 1961, (Washington, D.C.: GPO, 1988), 698-700.

33. Karnow, 310.

34. Robert Thompson in particular took this line and had no qualms at all about dealing with people like President Diem.

35. H.A.F. Hohler to Lord Home in an aforementioned letter *vis-à-vis* the war—"I do not consider . . . that senior American commanders in Viet Nam really understand its complexity". See Hohler to Home, 30 Jan. 1963, PRO, FO Vietnam/ 371/170100.

36. No less than three national leadership changes—punctuated by abortive coups—took place in the two years after the assassination. Generals Minh, Khanh, and Ky followed in rapid succession. For a detailed account of this period see Karnow, 335-385.

37. Unsigned memoranda, "Future of the British Advisory Mission in Viet-Nam (BRIAM)", 14 Oct. 1964, BRIAM (i.e., FO) Vietnam/371/175484.

38. Hickey, 80.

39. They felt that the Americans had made too much out of the guerilla versus guerilla concept. Similarly, the concept of counterinsurgency in which "The Army counters the Vietcong regular units; the Civil Guard counters the territorial units; the SDC and Hamlet Militia counter the village guerillas; political cadres counter political cadres; aid counters promises; and so on." was thought to beget a civil war situation in which neither side could claim to be a governing authority. These points are made by Robert Thompson in a situation report which is not dated other than upon receipt at the Public Records Office. However, analysis of the content reveals that it was authored some months after President Diem's assassination and in 1964. See Robert Thompson to the Foreign Office, 1964, PRO, FO Vietnam/371/175482/DV 1017/17.

40. That they were well-received is only too obvious. The BRIAM imprimatur was all over the GVN apparatus in the form of bureaucratic/organizational schema paralleling that employed in Malaya.

41. USOM was the civilian advisory organization in Vietnam that handled economic development assistance, as well as the CIA and Michigan State University endeavors. It thus served as a counterpart to MAAG and carried a lot of weight.

42. The Americans had already made clear to the British they expected to be the senior

partner in the Vietnam venture. So, they readily took the bait when the British from all outward appearances shifted their position ever so slightly. Even today there is great reluctance on the part of those who were present to admit that they should have relied more heavily upon the wisdom of British jungle experience. Frederick Nolting, the United States Ambassador to Diem's government, states in his memoirs—in a claim which is not well supported by archival material—that United States officials worked very closely with R.G.K. Thompson. See Nolting, 37. Whereas in his professional correspondence during the war he bluntly states "Regret . . . Thompson mission is badly off rails from standpoint US-UK coordination and that Thompson recommendations to GVN, whatever may be their intrinsic merit, are bound to complicate our task of bringing about essential reforms in GVN military and administrative structure. . . . Difficulties with Thompson Mission have been procedural and substantive." See United States Department of State, *Foreign Relations of the United States, 1961-1963*, Volume 1, Vietnam 1961 (Washington, D.C.:GPO, 1988), 698. Similar *post facto* misrepresentations are also made by Norman B. Hannah—also of the State Department—although Hannah concedes there were differences of opinion about how the war itself should have been prosecuted. See Norman B. Hannah, *The Key To Failure* (Lanham, MD.: Madison Books, 1987), 195.

43. A total of four police advisors were appointed in place of the original BRIAM staff. They had two-year contracts which could be terminated on a month's notice and were thus free to quit should the situation collapse. Similarly, Robert Thompson withheld his own intention to withdraw from Vietnam and phase out BRIAM while these arrangements were put into place. This information is revealed in a report titled "Viet-Nam: Appointment of Four Police Advisers in Place of BRIAM" from Saigon signed by Sir H. Caccia, D. F. Murray and E. H. Peck. See Caccia, Murray and Peck to the Foreign Office, 16 Oct. 1964, PRO, FO Vietnam/371/175484.

44. See the aforementioned unsigned memoranda, "Future of the British Advisory Mission in Viet-Nam (BRIAM)", 14 Oct. 1964, BRIAM (i.e., FO) Vietnam/371/175484.

45. This comment is made in a memorandum from the Secretary of State's Special Assistant for Vietnam (Sullivan) to the President's Special Assistant for National Security Affairs (Bundy). See, United States Department of State, *Foreign Relations of the United States, 1964-1968*, Volume 1, Vietnam 1964, (Washington, D.C.: GPO, 1992), 526.

46. The idea is floated in various memorandums. See, United States Department of State, *Foreign Relations of the United States, 1964-1968*, Volume 1, (Washington, D.C.: GPO, 1992), 79; United States Department of State, *Foreign Relations of the United States 1964-1968*, Volume 1, Vietnam 1964, (Washington, D.C.: GPO, 1992), 361. On the other hand, the noted French observer—Bernard Fall—of both French and American military endeavors in Vietnam deplored any American deference to the British experience in jungle warfare. In conversations with friends and colleagues he often liked to say the only thing Malaya and Vietnam had in common was their hot climate. Fall favored an approach the French called gridding (quadrillage) but acknowledged it required far more manpower than what was available in Vietnam. He said, "One doesn't start from the center of something and work one's way out, but he starts from the periphery and works one's way in . . . if it is done right, and if it is done in enough places at once, some communist units will finally get fixed . . . and caught." For just one of many very interesting books about Vietnam by this author see Bernard B. Fall, *Last Reflections on a War* (Garden City: Doubleday, 1967).

47. The men R.O.D. Noone worked with on the ground are said to have all been very professional advisers/combat specialists. Hence, there was a good deal of mutual respect and cooperation among colleagues unlike at higher levels in the chain of command. These men did not go out into the field with Noone and his teams and it is a shame such model

interaction was limited to planning and logistics at the lower levels of the chain of command.

48. See United States Department of State, *Foreign Relations of the United States, 1961-1963*, Volume 1, Vietnam 1961 (Washington, D.C.: GPO, 1991), 698, 699.

49. To quote the Ambassador "I sense nevertheless that Thompson is thoroughly annoyed." See, United States Department of State, *Foreign Relations of the United States*, Volume 1, Vietnam 1961, (Washington, D.C.: GPO, 1988), 700.

50. In the aforementioned cover letter and attached summary of a September 12, 1963 meeting, between Robert Thompson and Ambassador Lodge, Thompson boils down the American dilemma and parrots Ambassador Lodge's plight—"What do we do Now?"—to F.A. Warner. See Thompson to Warner, 18 Sep. 1963, PRO, BRIAM (i.e., FO) Vietnam/371/170102.

51. Elements of the Australian SAS based out of Kurmunda are reputed to have remained active in the Delta region of South Vietnam but otherwise most Commonwealth forces available in the region were fully committed to Borneo.

52. Similarly, individual operatives like R.O.D. Noone, Mike Jones and others continued to assist by other means in the counterinsurgency program from Thailand and South Vietnam for some years to come. However, this is another story and someday it may be told.

53. Training in intelligence, riot control and administration was also made available by the Special Branch and the Royal Malaysia Police in Kuala Lumpur. A summary of these activities is available in correspondence (letter and appendix) between Robert Thompson and A. J. Williams at the Foreign Office. See Thompson to Williams, 22 Aug. 1963, PRO, FO Vietnam/371/170101.

54. One of two wings at the school was devoted to the Vietnamese and Thais. John P. Cross served as the last Commandant of the JWS and states courses for the ARVN were an "open secret" despite British policy-maker's claim to pursue a policy of non-intervention towards South Vietnam. See Cross *A Face*, 207, 208.

55. This is essentially the argument made recently by Lind Michael with respect to America's involvement in Vietnam. For an interesting analysis of the Vietnam war's greater meaning see Michael Lind, *Vietnam, The Necessary War: A Reinterpretation of America's Most Disastrous Military Conflict*. (New York: Free Press 1999).

56. The Philippines made claims upon parts of Sabah and were quite vocal in their own protestations to the Malays about this point. However, they stopped short of becoming militarily involved in the conflict. Thus, it remained a war between Malaya/Britain and Indonesia and narrowly escaped taking on a greater regional dimension.

57. The fighting in of itself was nothing new. Various Dutch, British, Portugese, Malay and Indonesian elements have a history of skirmishes, raiding and periods of protracted warfare under their belts. Now, however, the remaining contestants were stronger and the stakes much greater. Concessions, provinces and trade issues were no longer at stake—countries themselves were on the table. In fact Sarawak, Brunei and Sabah were administered and referred to as three different countries by the British.

PART VI

The Indonesian Confrontation on Borneo and Its Spread to Peninsular Malaysia

The Senoi Praaq upon their departure for Vietnam left behind a war torn "Malaya" weakened by the perils of the Emergency but they returned a few long months later to find a bold new "Federation of Malaysia" in her place. The Emergency remained in remission, economic recovery was underway and promising new economic development schemes were already on the table. However, all was not well at home as Malaysia and Great Britain were now engaged in a battle with Indonesia over control of the new Malaysian states of Sabah and Sarawak (East Malaysia) and the Sultanate of Brunei on the island of Borneo.

The conflict, a series of stiff clashes the GOM dubbed as the Indonesian Confrontation, started in Brunei on December 8, 1962 and quickly spread first to Sabah and Sarawak and then to the Malaysian mainland itself. Nearly five years of hostilities would pass between the antagonists before the violence petered out. The founding of the Association of Southeast Asian Nations (ASEAN) in 1967 formalized an end to the trouble and brought about a rapprochement among the new nation states situated in Southeast Asia's lower rim.

The fighting initially involved vicious low intensity raids carried out for the most part by a hodgepodge of Indonesian communists and irregulars. Later small groups of regular army and some elite troops took part in larger scale incursions and this added fuel to the fire. For the Senoi Praaq the second of two guerilla wars had simply given way to a third. However, Vietnam aside, the essential difference between the Indonesian Confrontation and the Malayan Emergency lay in the fact that the Malayan communists were largely an isolated guerilla movement with no backers in a formal diplomatic sense. They terrorized the country over a long period of time but there was a built-in limit to their ability to intensify the struggle. On the other hand, Indonesia—a sovereign country and a foe with superior military power—was, in a more immediate sense, an elephantine threat to Malaysia. She could have raised the stakes and mobilized for a full blown war rather than try and engineer Malaysia's internal collapse through limited probing.

Hence, in yet another happy coincidence the unit was to profit from the misfortune of others and ergo a permanent place in the Malaysian order of battle was now in the offing. Malaysia in her latest hour of need would be very generous indeed with those who came to her assistance. And so, a Senoi Praaq cadre once again managed to arrive in the right place and at the right time to guarantee survival of the Senoi Praaq as a whole. At this juncture in the organization's evolution the Senoi Praaq thus engages in a rescue and salvage operation on Borneo, resumes operations on the peninsula and reestablishes in the eyes of the Malaysian government its *bona fides* as a combat unit.

In sum, the unit seeds itself on its own turf close to home, broadens its recruitment base and effects a lasting political rehabilitation. Key changes in the Senoi Praaq command structure are also in the offing and the Senoi Praaq's ability to adapt and replicate itself in form if not function becomes manifest.

Chapter 18

The Borneo Crisis

The Indonesian Confrontation from day one had the potential to surpass the Malayan Emergency in intensity and events might well have taken a turn in Indonesia's favor. Fortunately, however, a quick military response by British and Malay troops combined with on and off again diplomatic efforts in Jakarta, Kuala Lumpur and London to telegraph to Indonesia the depth of British/Malay resolve. The Borneo incursions were contained and on August 31, 1963, Malaya went forward with the formation of the Federation of Malaysia which incorporated Singapore, Sabah and Sarawak into her territorial mass. Just like that a new national entity was born in the heat of battle despite the Indonesian threats. Every prospect existed for Malaysia's retention of the Borneo states and a successful resolution of the conflict. However, in a geo-political sense the stakes were now much higher as Malaysian territorial integrity was at stake and there was thus an even greater reason to thwart the Indonesians.

GEOPOLITICAL DANGERS

Great Britain would have to avoid the perils of escalation on Borneo. British theater commanders were under great pressure to keep a lid on the situation through the use of force without provoking an endless sequence of tit for tat responses by their Indonesian adversaries; otherwise, matters might get out of hand and escalate into Britain's own version of the Vietnam War in the Far East—a politically unacceptable course of action during the very time she focused on the decolonization of her Empire. Nor were her commanders given *carte blanche* to go in and deliver a death blow, a deed which would be very costly and risk sending the wrong signals beyond regional boundaries.

Thus, overkill offered a dangerous and potentially disastrous solution to the problem at hand as did too insipid a response to the threat posed by the Indonesian incursions. The use of force had to be judicious and non-provocative but sufficient to neutralize the various elements of an enemy ensemble as they arrived on station along the Kalimantan border. An Indonesian blow had to be anticipated and blunted without provoking a rabid counter riposte, and well before Sukarno's men could achieve critical mass in any one particular place.

Numerous units were deployed along the border towards this end but they were insufficient in number to make a real difference should Indonesia decide to press her advantage. British and Malay forces were thus vulnerable at any given point to Indonesian military action on the lengthy frontier. There were simply not enough men and helicopters in place let alone the heavy fire power needed to counter concerted thrusts in a timely manner. As a consequence the British and Malays had to walk a fine line between losing control over the situation and becoming Sukarno's patsies lest he be afforded an opportunity to whip the Indonesian people into any more of a lather than he already had.

In many respects it was only a matter of time before regular Indonesian forces—who had begun to assume responsibility for the fighting from mercenaries and rebels—swamped the frontier area and broke through to the coast. Such a move would then allow them to link up with communist guerillas operating in these areas and overwhelm any remaining government units willing to fight onwards. Yes, there were communist rebels just about everywhere in Southeast Asia at that time and Malaya inclusive of the Borneo states and Indonesia were not exceptions.

The Borneo communists were—like their counterparts on the mainland—also Chinese dominated. The British dubbed the Borneo equivalent of the MCP, the Clandestine Communist Organization (CCO). However, they were not quite in the same league as the well-organized Malayan communists and they posed a lessor liability. That the two communist groups never managed to establish a substantial rapport of their own pointed up deficiencies sufficient to compromise the use of their shared ideology as a framework for pan-Asian subversion.[1] Nonetheless, the Borneo communists were actively courted by Sukarno's pro-communist government and a junction between these two threats whereby multiple lessor enemies might coalesce into a mortal danger had to be stopped at all costs.

THE ANSWER ALWAYS LIES IN INTELLIGENCE

What to do? The answer to this dilemma once again lay in the availability of timely intelligence whereby Great Britain and her Malaysian allies might accurately concentrate their own available men and material in preparation to strike decisive blows. There was no other way to break up and repel an invading force short of a major commitment of ground troops. But the requisite intelligence in these situations was never easy to come by and its acquisition on Borneo was made even more difficult by numerous exigent circumstances—geographic, demographic and political to name a few.

Most of Borneo is mountainous and covered by luscious jungle just as dense, if not more so in places, than that encountered on peninsular Malaysia proper. Thus, the going was much more difficult for combatants engaged in this theater even though they were already accustomed to jungle fighting. To complicate matters further Borneo was a map maker's nightmare and very little hard data was available about geographic conditions in the interior. Parts of the Malay peninsula towards the Thai border had never been mapped either but Borneo from a cartographic standpoint was even less well-understood than Malaya.

It would require more than simple improvisation to overcome this handicap before real manoeuver elements could operate with any degree of exactitude. Demographics also presented a formidable problem. The ethnic variety to be found among Borneo's indigenous population was most perplexing. There were Dayaks, Dusuns, Ibans, Kayans, Kelabits, Kenyahs, Punans, and Muruts to name a few of the twelve odd tribes to be communicated with and understood before anything else could be done. These peoples were distinct from one another and many lived in the operations area along the border between Sarawak, Sabah and Kalimantan. Thus, the linguistic aspects of opening a communications channel among the various parties alone presented as serious an impediment as did geography to the defense of the area.

Then, even if geographic and demographic concerns could be overcome there was the ever present political angle to be negotiated. How to make the tribes work together and with the Malays towards a common good? Much ado about nothing is made over supposed Malay/Orang Asli friction but the Borneo tribes were another matter altogether. Not only do they not take too well to the typical Malay but quite a few of these peoples also had longstanding feuds with one another which make today's racial problems in the Balkans seem rather tame but for their most recent and flagrant eruption. Obviously, the means whereby these different groups might be manipulated along military lines in the pursuit of mutual self interest was less than obvious.

THE WANTING SOLUTION

The most realistic immediate solution to this dilemma lay in the use of special forces and the creation of a local levy along the lines of the original Senoi Praaq. A similar scenario had worked well on the peninsula and as we know there is a tendency for people to gravitate towards the familiar in times of crisis. Not surprisingly, General Sir Walter Walker—the new theater commander for Borneo—would welcome overtures from special forces personnel. In this regard, the Special Air Service itself was—after some smooth talking at the Defense Ministry in London—already back on scene and in action along the border with Kalimantan. However, its troopers were stretched too thin to make an effective screen.[2]

The area in question was so vast the SAS troopers alone could not monitor it even if they were at full strength—which they were not. As a consequence only the most likely approaches across the 970 mile land frontier were subject to surveillance and there were no men to spare for offensive operations.[3] The SAS back in Britain had been subjected to the budget cutter's scalpel and had fallen on hard times not long after their departure from Malaya and a short assignment in Oman.

Consequently, they were under strength and one did not easily recruit these people on street corners.[4] Be this as it may, the war on Borneo was a godsend for them and a natural outlet for their talents and energies but it would take time to rebuild their numbers.

In the interim the SAS teams were vulnerable to the enemy and they had to take some casualties in an environment in which the term 'Missing in Action' was not applicable. One either made it back to base or was written off as dead in no uncertain terms and under the worst possible circumstances. A merciful death amidst headhunters, dreadful tropical diseases and venomous serpents galore, not to mention the Indonesian enemy, was just about the only element not in short supply during the Confrontation. It virtually goes without saying that most of the fighters running around on Borneo had never heard of the Geneva Convention.

Capture for the unlucky who were not killed outright more often than not, as during the Emergency and in Vietnam, meant a tortured end. Goodness knows what sort of refinements the Indonesian intruders put into place besides the usual fare. Friends and relatives of the victims at best could take cold comfort in the knowledge that their loved ones did not suffer long. Indonesian methods were reported to be very crude and gruesome and we may surmise prisoners usually died from shock before they could be broken and divulge any information.[5] What a pity the purpose of unpleasantness so often is sport and not true intelligence collection. Barbarity was never easy to deal with and obviously the SAS troopers needed all the help they could get. Not surprisingly they chose to look to locally recruited assistance and saw the establishment of another irregular "scout" type unit as an option. Unfortunately, nothing was ever easy—especially on Borneo—and they experienced some difficulty in finding a suitable formula for such a unit.

The immediate question became why not reestablish contact with the Sarawak Rangers and recruit their Iban headhunters who had heretofore worked with the SAS on the Malayan mainland? The SAS regrettably had experienced difficulty with some members of this unit, and moreover the Malaysian government was averse to providing them with automatic weapons in any great quantity. In sum their on-the-job performance during the Emergency left something to be desired and this time around the new Malaysian government might find it difficult to disarm them on Borneo.[6] Nonetheless, the SAS did in fact look up a number of former ranger personnel and found that most of them had in fact become taxi drivers in Kuching and so the legend of the Kuching taxi drivers became true but only in a *post facto* sense. These men albeit more or less suitable material at one time were now deemed to be out of touch with the jungle and bereft of an edge.[7]

Chapter 19

A Senoi Praaq Salvage Operation

Not only did the British on Borneo run great risk at the hands of the Indonesians but they also very nearly forged a defective unit more dangerous to themselves than the stated enemy. Their confidence was misplaced through reliance on a generic Borneo levy known as the Borneo Border Scouts (BBS) who were proven, most decidedly, not to be on par with the original Senoi Praaq. Nor did they stack up well in comparative terms against some of the Montagnards the Senoi Praaq had recruited previously in Vietnam.

The BBS were incapable of withstanding pressure and thus they were more of a liability than an asset to the common cause. Bad troops like bad weapons, friends and women will always let you down and or run away with someone else. This turned out to be the BBS case in so much as the scouts suffered from a severe case of the "evil too(s)": they were levied from too great an array of peoples of suspect loyalty; their training was too conventional; and it was too hastily undertaken.

AN ABORTIVE FIRST TRY: THE BORNEO BORDER SCOUTS FIZZLE

There has been sort of a juxtaposition over time of the character traits exhibited by the peninsular Orang Asli and the tribal peoples of Borneo. The peninsular Orang Asli are often depicted by academic researchers as being meek and non-aggressive individuals whereas the Borneo peoples are promoted in the popular literature as being aggressive and quite mercurial. In point of fact the war like capabilities of the Borneo tribes people are somewhat overrated and their reputation as fierce head hunters overstated.

True a lot of them took skulls as trophies, while some worshiped a war god and/or were known to practice human sacrifice but their socio-cultural proclivities are misleading. The acquisition of skull trophies typically did not occur after battles but as a result of singular murders of unwary individuals conducted on the sly between members of rival tribal groups.[8] As for human sacrifice and war god worship there is limited evidence of these practices and they appear much less common in the greater scheme of things.[9] And so, the pell-mell exhibition of such traits across a spectrum of Borneo peoples did not translate into an available pool of individuals capable of being transformed into the perfect island-wide warrior class.

The Border Scouts were also poorly trained. SAS personnel were too busy at the time with operational matters to train thoroughly the new unit and essentially left this critical task to the Gurkhas. True, the Gurkhas were a heavy contact, they made excellent shock troops, and they were suitable for most assignments. Unfortunately, jungle warfare—from the standpoint of irregular operations conducted by men born to the ways of the jungle—was not their strong suit although they did not do too badly during the Emergency.[10] Gurkha elements as we know were very useful early on in breaking up large MRLA troop concentrations in the tri-state area of Negri Sembilan, Pahang and Selangor but only late in the day and in limited fashion did they dabble in real guerilla warfare.

In this respect the Gurkhas attempted to move in on Senoi Praaq turf in northern Malaya; however, they had trouble operating against small MRLA units and lacked the skills to ferret out intelligence on their own. Their physical appearance (i.e., Nepali) denied them freedom of movement and as a consequence they were tethered to the Special Branch. As foreigners their presence in Orang Asli areas was easily spotted and information about their location was quickly relayed over the bush telegraph to the communists. This suggests the Gurkha's method of operations was somewhat rigid and a little bit crude in comparison to the Senoi Praaq but it was no less deadly if they could make contact.[11]

However, despite the shortcomings of their operations in a guerilla environment the results the Gurkhas managed to generate were sufficient for a second try on Borneo. Senoi Praaq forces, whether in the doghouse back on the peninsula or otherwise engaged in Vietnam, were dispersed and unavailable in the short term for such work. Hence, the Gurkhas were enlisted to train and to provide—through seconded Gurkha NCOs—section level leadership to the BBS on a fair scale.[12]

In a major effort to mobilize the indigenous population about a thousand Borneo tribesmen were induced to join the new unit and placed under arms. Unfortunately, not only were the wrong people selected and instructed in the wrong tactics but they were also trained for only three weeks. This accelerated training schedule was far too short a time to impart the necessary skills and it was not comparable to the three months the SAS spent on the Senoi Praaq. Such skimpy preparation could not be helped as British and Malay forces were under intense pressure from the Indonesians; however, the botched training soon turned out to be the telltale agent of defeat for the BBS and Gurkhas. Their failure in one of the first encounters by British and Malay forces with Indonesian regulars brought about a shakeup of the BBS and a reorientation of its mission.

LONG JAWI BLOODBATH

The initial BBS levy largely based upon Sarawak tribal peoples was dispersed throughout numerous posts in the vicinity of the administrative town of Belaga. Belaga is located in a forward area not far from the Sabah and Kalimantan borders, and like most posts it was dangerously exposed.[13] Its retention was an integral part of the strategy formulated by the British and the Malays in which their joint forces fought as far forward as possible. Their objective was to prevent the Indonesians from driving the civilian population away from the border area and thus to prevent the subversion of the inhabitants and the forfeit of their loyalty to Malaysia.[14] Consequently, to their way of thinking the battle would be won or lost at the border and certain perilous dangers had to be born with a stiff upper lip.

The village of Long Jawi was one of these danger spots and its fall on September 28th set off the alarm bells as of late 1963. The post protecting this particular village was manned by a large contingent of Border Scouts, their Gurkha handlers and assorted Police Field Force radio technicians. British troops also rotated through the area periodically and it should never have been seized, pillaged and put to the torch so easily. In this business though there was not safety in numbers and the Indonesian forces were able to infiltrate the camp and spy out the defenses due to deficiencies in the BBS discipline structure.

Indonesian troops were so effective that members of their reconnaissance party were even mistaken for local tribes people during a general town meeting and treated to a complimentary briefing by British officers. A bloodbath ensued as a result and to make a long story short the village was overrun after heavy fighting by a company strength unit of Indonesian regulars. Members of the garrison held out for a time against overwhelming odds but gradually many of the border scouts began to break under the strain and fade away. The remainder of the troops who opted to stay and fight failed to break contact once it became clear further resistence was futile and they were overrun. Only three men out of more than two dozen were known to have survived.[15]

As a consequence of the Long Jawi fiasco, the indefatigable Gurkha leader John P. Cross was eventually relieved of command. Colonel Cross, an extremely non-conventional officer, was a controversial figure who rubbed a lot of eminent personages the wrong way and who was known to call a spade a spade. Some acquaintances said he had a "wild hair" up his ass.[16] Be this as it may he was also new to the job—having just been called back from leave in Britain—and had not had time to take the spit and polish out of BBS training. His efforts to improve upon the BBS formula were laudable but to no avail. Cross could not bend time to his will and events overtook his men. Most of the changes he had in mind could not be effected immediately and little to no progress came about prior to Long Jawi.[17]

THE SENOI PRAAQ TAKEOVER

The BBS failure was as much administrative nonperformance as it was a military bummer. The fault lay not at the feet of any one individual. Ultimately though, the Supreme Command saw fit to turn over the unit to the Senoi Praaq proper. Shortly after their return from Vietnam a *defacto* handover took place in which Richard Noone and his own controversial deputy—Norman Herbolt—gained control of the BBS from a disgruntled John P. Cross. It was an icy meeting in which Cross meted out glacial silence to Herbolt. Later he would look back on the negotiations and refer to the deputy Senoi Praaq commander as " . . . the odious Norman Herbolt."[18]

When all was said and done R.O.D. Noone and Norman Herbolt would also have control over a second subsidiary organization—the newly established Sabah Border Scouts (SBS), which was really an extension of the BBS to which John P. Cross had acquired titular control not long after Long Jawi. This added another 1000 men into the mix for a total of 2500.[19]

Initially the BBS referred technically to several miscellaneous detachments of Dayaks and Ibans trained and operated by British military units in Sarawak on a rather ad hoc basis. The Senoi Praaq element per se did not emerge until May, 1964 when R.O.D. Noone finalized the set up of the SBS (using mainly Murut tribesmen) headquartered at Keningau in Sabah.

This changing of the guard represents a fine example of the ever present friction which underlies the relationship between regular military and paramilitary leaders. In this particular instance exigencies of the moment prescribed that the traditional military dominance over operations and prejudice against the special forces be cast aside. The fateful decision which had such great impact upon the Senoi Praaq future was made March 8, 1964, at the Kuala Lumpur offices of Sir Claude Fenner, the Inspector General of Police (IGP). The Inspector General, dubbed "King Claude" by his staff, had a lot of clout and was able to pull off such a move without too much opposition from the regular military thanks to his forceful personality. Other parties in the know at the time were less complimentary of his merits and quick to describe Sir Claude as a "shit" in more general terms as opposed to a forceful prescient director.[20]

That Richard Noone managed to remain above much of this intrigue and mud slinging is testimony to his prowess both as an officer and a gentleman.[21] John P. Cross refrained in his memoirs—which were laden with abuse for a number of people—from making any derogatory comments whatsoever about this particular individual. And so, the Senoi Praaq were soon to be back in operation on a permanent basis but first they had to perform a small goodwill gesture towards the government and sort things out on Borneo.

Chapter 20

The Senoi Praaq Prescription

Subsequent to a further meeting between Colonels Noone and Cross in Brunei a short time later—to work out details of the handover—the BBS portfolio (now inclusive of the SBS) came under exclusive Senoi Praaq control. For this new assignment—aimed at BBS reorganization, which was to be very much like the one they had just experienced in Vietnam—Richard Noone once again took along his favorite operatives. The key personnel involved—Charles H. Ley, Zainul Alsagoff and the ubiquitous gunman Bah Tangga—had all been with him in Vietnam and by this point everyone knew their role by heart.

THE NEW GIG

To round off the contingent R.O.D. Noone even had his old side kick—the Senoi Praaq *eminance grise*—Norman Herbolt in tow and thus a complete team with whom to work. Together they wasted no time in taking control over twenty-five BBS camps and soon had their new charges in hand. The unit was revamped into a full-service combat/tracking organization within a few short months, primed and ready for action.

From an operational standpoint BBS status as auxiliary police did not change and eventually they too were made a part of the General Operations Force (GOF). However, significant adjustments were made in the existent training schedule and mission orientation. The training cycle was extended, intensified, focused to a greater extent upon sleuthing as opposed to combat and thus modified to meet existing needs more realistically. As a result the conventional military side of the operation gave way to a more casual paramilitary orientation which elevated the importance of the intelligence function.

John P. Cross had in fact recognized the limitations of an overly orthodox formation, and he is on record as having made a number of suggestions to improve BBS performance.[22] Thus as a practical matter Noone, et al simply expedited "improvements" recommended earlier to General Walker. True, they were very much the government's specialists in guerilla warfare but their political connections turned out to be much better than those of Colonel Cross. By the time of R.O.D. Noone's involvement in the Borneo campaign George Lea had become General Sir George Lea KCB, DSO, MBE and took over from General Walker as Director of Operations. One of R.O.D. Noone's titles on Borneo was tribal advisor to the Director of Operations. Thus, the Senoi Praaq enjoyed superior political fortunes and happened to have a *modus operandi* that obviously traveled very well. In sum, a fortuitous and an unbeatable combination of circumstances allowed them to assume/usurp control over the operation.

Suffice it to say casual *post facto* observations about unit orientation and the political environment of the day—despite early Border Scout setbacks—must in no way be interpreted to impugn the Gurkhas or their leadership. Neither John P. Cross nor his men failed in their assignment among the Border Scouts on Borneo. Instead, R.O.D. Noone and his cohorts moved in on this operation, built upon the foundations laid by their illustrious predecessor and forged the BBS into a more functional unit cast in a Senoi Praaq as opposed to a Gurkha image.

At this point the Senoi Praaq matured into a counterinsurgency franchise of sorts in the sense that its BBS proteges managed to develop into a pseudo variation of the original Senoi Praaq theme which had been adapted from early SAS operations on peninsular Malaya. The Senoi Praaq's artful rehabilitation of the BBS hence presaged its own assumption of the SAS training function for a cross section of special forces in Malaysia. As a consequence of this feat it was about to become, in a roundabout way, victim—albeit willing—of its own success.

CLARET OPERATIONS

The Indonesian forces employed additional conventional troops to continue their bloody assaults throughout 1963 and into 1964. In yet another effort to intimidate the Malaysian government Royal Malay Regiment (RMR) elements located in the Tawau Residency of Sabah at Kalabakan were targeted and singled out for a real beating. The RMR's heavy losses sent shock waves once again through the GOM. The Indonesian Confrontation had thus progressed from a series of armed assaults to an undeclared war of sorts in which the regular Indonesian military took over the dominant role from proxy forces and mercenaries.

Every reasonable diplomatic effort was made to come to terms with the Indonesians through a new round of negotiations but these peace talks collapsed in June 1964, signaling the need for extreme measures. It was time to do more than garrison and defend the border areas in a static mode. Thus began a sequence of counter strokes known collectively as "Claret" operations and the decision made to take the fight to the enemy.

Claret operations were designed to drive the enemy away from the immediate border area and hence to prevent any incursions into Sabah and Sarawak. These preemptive strikes were implemented in sequential phases: The river routes Indonesian troops used to approach Malaysian territory were first identified, then their main camps were located and their force strength estimated. The idea was to scout out the Kalimantan side of the border and extract the intelligence which British commanders needed to beef up their "kill units" and vector them into contact with inferior Indonesian forces.

The essential division of labor was such that the Gurkhas supplied men for the assault groups and the BBS and SAS provided the scouts who actually led the missions. The BBS now served as a general purpose reservoir from which the SAS drew select scouts for participation in their forays into Kalimantan. R.O.D. Noone placed his best specimens at the disposal of Major Jon Edwards—a snake eater in the truest but non literal sense of the term—for this purpose.[23] Major Edwards in turn parceled out these "cross border scouts" over a two year period between 1964-1966 to various SAS teams.[24]

Their sorties were sanguine affairs in which blood flowed freely—as the name for the operations implies—backed up by 155 millimeter artillery. Air support was also available but employed with discretion on an as needed basis only in extreme emergencies. Thus, parsimony was in vogue and gratuitous dependence upon heavy firepower in disfavor in keeping with the doctrine that overkill is not always good for you as it does not necessarily translate into success in these situations.

In sum Claret operations were a carefully controlled affair designed to bleed the enemy and weaken his resolve. The results of this particular bid at comparative advantage were mixed, however. The desired objective was accomplished on Borneo. Indonesian troops were induced to stay on their side of the border and they ceased their incursions into Sabah and Sarawak. But the enemy proved to be a little more tenacious and resourceful than expected and unfortunately, for Malaysia, Government of Indonesia (GOI) in response to the Claret operations shifted the onus of the attacks to the peninsula itself.

MORE DARK DAYS AS THE CONFRONTATION SPREADS TO MALAYSIA'S MAINLAND

With the fighting on Borneo at a dead end GOI decided to up the ante and break the stalemate by targeting directly Malaysia's mainland. It dispatched infiltrators, saboteurs, and regular troops alike in several large landing parties. Singapore island also became a focal point of this activity. A Shell Oil Corporation facility, Merdeka bridge, the Singapore water supply line from Johore and the Singapore-Kuala Lumpur express railroad line were hit in June and July, 1964.[25] By August the focus of the pressure upon GOM thus shifted much closer to home. The objective was the destabilization of Malaysia itself.

This strategy was carefully thought out and a viable one as West Malaysia was by no means a stable political entity. There were a variety of fault lines in the national identity to be exploited independent of the recent Chinese based communist insurrection. Hence, a new phase in the fighting began in mid-to-late1964 with numerous small-scale insertions of seaborne raiders from the Korps Komando Operasi (KKO) along the west coast of Malaysia's mainland.[26]

These initial probes were followed by two robust attacks by sea and air upon Johore state—at Pontian and Labis—which are understood to have been loosely coordinated.[27] Numerous smaller raids followed over the next two years. GOI once again used regular Indonesian troops supplemented by Malaysian exiles and resident dissidents to mount these assaults, quite a few of which were either intercepted at sea or otherwise foiled once they had disgorged their cargos. A front organization—the Tentara Nasional Malaya (TNM)—similar to the NKNA was created to lend legitimacy to these activities and to disguise thinly GOI involvement. The TNM served as an umbrella organization for disaffected Malays and Chinese opposed to the regime of Tunku Abdul Rahman, Malaysia's first prime minister.

THE INITIAL INDONESIAN ASSAULT AT PONTAIN

Large pockets of Javanese immigrants lived in Johore state and GOI initially attempted to mobilize them against GOM. The first major cluster of armed assaults took place near Pontian—the closest point on the peninsula to Indonesia—in August 1964 when several seaborne parties placed at approximately 100 men came ashore about 30 miles north of the Straits of Johore. The Mangrove swamps blanketing this region provided ideal conditions for concealment and it became a favorite landing area for the Indonesians over the next two years. However, these initial raiders did not make much headway; they failed to generate support among the local people and the Malaysian security forces dealt with them before a toehold could be achieved and a provisional government established.

GOM committed a number of units to this operation including a company of the RMR, a reconnaissance regiment, a Senoi Praaq squadron and various Police Field Force personnel. These GOM forces quickly recovered substantial quantities of arms, ammunition and equipment.

The Senoi Praaq were employed at Pontian on a trial basis in order to gauge their utility in lowland areas and they fell straight back into stride. Half the raiders involved were captured shortly after they landed and the others apprehended over a period of several months. Typically, these captures came about—in a repeat of the pattern the Senoi Praaq established in the Malayan highlands—once the insurgents broke cover to approach local villagers for food or otherwise left their tracks behind. After these first encounters with the enemy the regular military forces cordoned off a 200 square mile area around Pontian. Senoi Praaq and Police Field Force units then set up checkpoints and swept the area until a smaller eight mile perimeter had been established. At this point hostile contacts multiplied and Indonesian resistance fizzled out shortly thereafter.

The effects of this initial raid on Pontian were more political than military and they were felt more so in Kuala Lumpur than in the countryside. Official sources cited the willingness of the local population to assist security forces and to provide them with good intelligence as the primary criteria explaining the government's success. However, the Indonesians would not be easily deterred; they changed their tactics and soon tried another large scale operation. One month later in September, 1964, they looked further inland and targeted the Labis area of north-central Johore. Substantial airborne forces were employed in this engagement. As with the Pontian incursions there was sound underlying logic behind this choice of operations area. In this particular location, a considerable Chinese population was present; the Labis region itself had long been a communist stronghold of some significance to the MCP; and the particular area flanked by the main range had remained a center of communist activity late into the Emergency.[28]

THE SENOI PRAAQ RESUME OPERATIONS IN FORCE: THE LABIS RAID

Unfortunately for the Indonesians the Labis operation also failed despite its plausible rationale. Numerous plane loads of men and material were dropped in the area during early September, 1964, but the intruders never managed to mass or form a base camp from which to operate. Nor did they succeed in rallying the local Chinese to their cause despite the fact that these people were among the last to be pacified in southern Malaya during the Emergency. Once again, sizable quantities of ammunition, food and medicine were seized by security forces. All but two men in this force were known to have been eliminated. Members of the Senoi Praaq present at the time reported to the author that this body of armed men broke up into clusters of soldiers who were eventually hunted down and taken out of the picture.[29]

Veterans of the Labis fighting have attributed the raid's failure to two key reasons—poor planning on the part of Indonesian military and the introduction of the Senoi Praaq into the fighting en group. The Indonesian troops dropped in the Labis region were haphazardly inserted and there was insufficient follow-up from a logistics standpoint. Similarly, Malaysian Security Force resources had been sufficiently taxed that GOM had no alternative but to lean more heavily on the Senoi Praaq. The unit thus made its first appearance en masse beyond the BOA at this juncture and it set up a temporary tactical headquarters in Johore.[30] This moment signals its formal rehabilitation and debut as a "Malaysian" national security instrument, as distinct from the freelance cadres that operated in Vietnam's central highlands and the coterie of Senoi Praaq who joined R.O.D. Noone in the defense of Sabah and Sarawak.[31]

THE SENOI PRAAQ IN TRANSITION: THE OLD HANDS DISPERSE

Now, the Senoi Praaq was no longer a private army under the command of R.O.D. Noone or his creature Norman Herbolt. Control of the organization had passed to one of his younger more innovative proteges, Mohamed Ruslan bin Iskandar Abdullah. Colonel Noone for his part was a confirmed war junkie and so he left Malaysia not long afterwards and made his way back onto the SEATO payroll as an advisor and planning officer posted to north-east Thailand. Hence, he continued to live the life until shortly before his death a few years later when he succumbed to natural causes brought on by years of hardship and self-destructive personal behavior.[32]

Noone would appear in Malaysia from time to time but this move marked a definitive change in his life and that of the plight of the Senoi Praaq. One of his more noteworthy reappearances on scene in Malaysia took place during the massive search for the presumed CIA operative—James H. W. Thompson of Thai Silk Company fame—who disappeared while on a visit to Malaysia's Cameron Highlands in 1967.[33] As a matter of professional courtesy Noone took a Senoi Praaq squadron back into hand and turned the place upside down looking—albeit unsuccessfully— for Jim Thompson who was thought to have met a sinister fate.[34]

Jim Thompson was not the only operative to disappear into thin air so to speak. Norman Herbolt also dropped out of sight shortly beforehand and he is thought possibly to have been killed on the sly.[35] He never gave up his own penchant for intrigue and he was involved in far too many black market deals of an unsavory nature. Poor Norman turned out to be just as unpopular on Borneo as he was on the peninsula and by this point in his career he was in that kind of awkward situation in which if the bad guys did not get you then the good guys surely would. There was no such thing as a disposal problem on Borneo or in the hinterlands of Malaysia for that matter and he should have been more prudent.[36] And so, operational control of the BBS per se eventually went to Charles Ley who later joined SEATO and shadowed R.O.D. Noone's career path for some years.[37]

THE COURSE OF THE FIGHTING

As for the fighting on the mainland itself, there is some discrepancy between the specific numbers of attacking Indonesian paratroopers (who went by the designation RPKAD) cited in published sources and the estimates quoted by Orang Asli present at the time. The number of drops executed by the Indonesian air force is also a matter for some speculation.[38] Figures commonly cited in the literature—which would appear to emanate from the same official sources—routinely mention two drops of approximately one hundred men in total.[39] However, the cargo capacity of a pair of C-130s, the type of plane employed, is somewhat greater even if provisions are made for materiel. Thus, either the planes were flying at less than capacity or the figures cited are conservative. The more likely explanation is that the numbers were kept small for psychological and political reasons. This is particularly so, given the dread and extreme pessimism the Pontian and Labis raids induced among officials at the Malaysian Ministry of Defense.

The conservative official reports may also be attributed to the tendency by both GOM and GOI to play down publicly their force strength. Neither side wanted to prompt the other to escalate the conflict into a full-blown war, but both wanted to win. However, despite the political obfuscation and vagary we do know that the raids were of sufficient magnitude to provoke a crisis in Kuala Lumpur. Malaysia declared a state of emergency shortly after the landings at Pontian and spent a nerve-wracking year. A wave of terrorist bombings struck Kuala Lumpur at about the same time and this complicated the situation immeasurably. Some stalwart European and Eurasian estate managers who stood fast during the Emergency began to transfer their wealth out of the country and a climate of fear prevailed.[40]

Thus, Indonesian pressure upon peninsular Malaysia continued well beyond the Pontian and Labis assaults throughout 1964 and well into 1965. On October 29, 1964 additional seaborne landings occurred in Malacca and Johore. These were followed by renewed attempts in November and December to infiltrate Singapore. The Senoi Praaq in several highly successful operations quickly put an end to these stabs aimed at Singapore. They in fact came to the aid of the Singapore Infantry Regiment after one of its patrols in Johore was ambushed and wiped out.[41] The Indonesians involved were tracked down, taken prisoner and turned over to the comrades of their victims. In this particular instance the prisoners were then marched off a short distance and presumably shot. Sounds of firing were heard and the men were never seen again. They were not kept for propaganda purposes.

British naval forces intercepted landings off of Cape Rachado on January 8, 1965 near the Malacca-Negri Sembilan border, and one day later they thwarted further efforts to come ashore in the vicinity of Kukup, Johore. Subsequent to these operations there were additional seaborne landings in May 1965. Also, the east coast of Johore in the vicinity of Tanjong Pen-Gelih was targeted and substantial Indonesian forces managed to entrench themselves briefly in old Japanese fortifications only to be dislodged by British air power before they could be reinforced.[42]

Senoi Praaq contingents were present for most if not all of these engagements and played no small part in the Indonesian defeat. Troops were lent typically to conventional forces on an as needed basis to conduct off road operations since these soldiers once removed from solid ground did not operate effectively. The unit was commended numerous times by British and Malay general officers for its signal role in the Confrontation's peninsula phase. This stellar performance more or less guaranteed the Senoi Praaq a permanent place in the Malaysian order of battle.

Part VI: Notes

1. The Indonesians are thought to have made an effort through their embassy in Bangkok to stir up the MRLA along the Thai border with Malaysia but for some reason a major thrust was not made southwards although the Indonesian Confrontation was exactly the type of occurrence one would expect to stimulate MCP activity. Instead the MRLA held back somewhat and did not resume a true offensive posture until 1969.

2. Tony Geraghty, *Who Dares Wins: The Story of the Special Air Service 1950-1982* (London: Fontana/Collins, 1983), 54, 55, 56.

3. Seven hundred miles of this front were actively monitored by the SAS with the help of local tribes people. See Geraghty, 56.

4. Two SAS squadrons were axed at the end of the Malayan Emergency. A squadron arrived on Borneo in January 1963 and went immediately into operation. See Geraghty, 55, 62, 63.

5. The SAS, according to an interviewee in a position to comment upon such matters, are not reported to have lost many men on Borneo. SAS casualties during the Borneo campaign were seven killed (three in one helicopter accident) and several wounded. However, the reader can be sure anyone unlucky enough to be captured died appallingly. Geraghty, 72-73. The Indonesians are also said to have been lazy and want to kill captives who could not march even if this implied shallow interrogation. Then they would brag about it to the locals. See Peter Dickens, *SAS : Secret War in South-East Asia: 22 Special Air Service Regiment in the Borneo Campaign, 1963-1966* (New York: Ivy Books, 1983), 91.

6. These difficulties have long since been resolved and the Sarawak Rangers expanded into the Malaysian Rangers as an integral part of the General Operations Force. As such they play their own central role in the overall balance of the power game which extends to Borneo. Similarly, there was also a great deal of resistance to the BBS concept by the colonial administrations of Sabah and Sarawak. The British expatriate community was suspicious and resistant to the idea of outsiders arming and taking over the leadership of indigenous troops. See John P. Cross, *Jungle Warfare Experiences and Encounters* (London: Arms and Armour Press, 1989), 158. John P. Cross, *A Face Like A Chicken's Backside: An Unconventional Soldier in South East Asia, 1948-1971* (London: Greenhill Books, 1996), 143.

7. See Geraghty, 55.

8. Cross, 146. This practice is now illegal and occurs with much less frequency but

obviously old habits die hard and even today any number of victims continue to be claimed. The author visited several Iban long houses and baskets full of skulls remain an integral part of the decor. Indeed, some looked fairly fresh and one wonders about their origin.

9. Cross, *A Face*, 166, 170.

10. The Gurkhas—who are best thought of as the fists of the government as opposed to its eyes and ears—thus imparted an orthodox infantry quality upon the BBS that was inappropriate for their intended role. See Geraghty, 61.

11. John P. Cross and his Gurkhas in particular are known to have engaged in a static form of reactive patrolling. They would lie in wait for an elusive prey weeks on end at a particular ambush site often to no avail and at great personal cost. After one such outing Gurkhas back at base thought he had been pushed too hard by command and aghast at his appearance even asked Cross "Sahib what have they done to you? . . . Why are they trying to kill you?" See Cross, *A Face*, 115, 119.

12. See Cross, *A Face*, 144.

13. Indonesian troops had—should they choose to engage this target—to travel only some fifty odd miles beyond Long Jawi via the Rejang River. Thus it was well within striking distance for a spoiling attack or a reconnaissance-in-force.

14. Cross, *A Face*, 143.

15. One Borneo Border Scout and two Gurkhas made good their escape and survived. It did not take the Indonesians long to round up the stray scouts. Most of these men were then bound, made into examples for the village people to see, and executed. The village itself was looted and burned to the ground. The fire must have made quite a sight as some of the native long houses on Borneo extended a half-mile in length and were home to more than one hundred families. See Geraghty, 62; Cross, *A Face*, 151-152, and Cross, *Jungle Warfare*, 163, 173. However, according to SAS sources official military figures sought to down play the Long Jawi fiasco and cite four killed (two Gurkha soldiers, one Border Scout and a policeman. In this instance John P. Cross's figures are thought to be more accurate and are mirrored by those of Geraghty. Be this as it may, the Gurkhas later took revenge and eliminated the attacking force as it attempted to return back across the border.

16. Nonetheless John P. Cross, at least initially, had powerful backers who also knew the jungle well. Major General Walter Walker—General Officer Commanding (GOC) 17th Gurkha division—personally chose Cross to fulfill this mission. By this point General Walker had become the Commander of British Forces in Borneo and could pick and choose his subordinates. See Harold James and Denis Sheil-Small, *The Undeclared War: The Story of the Indonesian Confrontation, 1962-1966* (Totowa, NJ: Rowman and Littlefield, 1971), 46, 68.

17. From Colonel Cross's memoirs—*A Face Like A Chicken's Backside: An Unconventional Soldier in South East Asia, 1948-1971*—one may gather that the intensity of the internal political battles—Malay/British, British/Commonwealth and British/British —was equal to the fury of their common struggle against the enemy. The book is laced with acidic albeit objective commentary about the bureaucratic infighting and prerogatives of the various parties involved. For a few examples that outline Cross's frustrations with bureaucracy and which date all the way back to his troubles in the BOA with the Senoi Praaq see Cross, *A Face*, 89, 102, 105, 108, 120, 124, 129, 144 and 176.

18. See Cross, *A Face*, 173.

19. Overall BBS force strength grew during Major Cross's tenure from a smallish unit, begun with something over 1000 men, to around 1500 personnel in Sarawak and a further 1000 based in Sabah under BBS auspices. So he really does deserve a fair amount of the credit for their later successes. See Cross, *A Face*, 143, 147. John P. Cross lasted on Borneo,

in a formal sense, from July 1963 to May 1964. May 14th 1964 is the date on his letter of resignation but he was notified informally by Tom Turnbill that he was no longer responsible for the BBS on April 10th 1964. He left Borneo in July 1964 for a well earned leave in Britain and then returned to duty with the Gurkhas on the peninsula. See Cross, *A Face* 178, 180.

20. Claude Fenner was a strong IGP, but much of the credit for the success of joint police/military operations in Sarawak (Borneo) should go to Roy Henry, Head of Special Branch 1963-1965 and then Commissioner of Police. An operational policeman in Malaya since 1952, he understood military needs but supported his own force. He went on to become Commissioner of Police in Fiji and then Hong Kong.

21. This observation about R.O.D. Noone's character has also been confirmed verbally to the author by numerous sources—British, Malay, and Orang Asli. To quote one of these people "if Dick couldn't do you a good turn, he certainly wouldn't do you a bad one."

22. See Cross, *A Face,* 144-147, 150, 152 and 157.

23. The author is told clarification is in order on this point. For the record, Jon Edwards has a great knowledge of snakes and recently mentioned to a close friend that he would never knowingly eat a snake, although he would eat anything else.This is because he loves them too much.

24. On some occasions he led them across the border himself, although this was against his brief. See Dickens, 154, 155, 219.

25. J.A.C. Mackie, *Konfrontasi: The Indonesia-Malaysia Dispute 1963-1966* (Kuala Lumpur: Oxford University Press, 1974), 258.

26. These landings were timed during September to correspond with a second wave of race riots in Singapore provoked in part by Indonesian provocateurs. Mackie, 259.

27. The general idea was for the Pontian group to rendezvous some distance inland at mount Pulai, establish a guerilla base, and then link up at a future point in time with the men dropped at Labis. See James and Sheil-Small, 146-148.

28. In a perfect world the various insurgent groups involved would have joined forces at some point and begun to train any locals who rallied to their cause. James and Sheil-Small, 146, 148.

29. Outright elimination was not always desirable, given the usual need to debrief prisoners and GOM's more novel desire to exert psychological pressure upon GOI through the trial and execution of prisoners. Indonesians captured on the mainland were not viewed as prisoners of war and thus they were considered fair game for the hangman's noose. Ultimately though this policy was discontinued due to the bad publicity it provoked.

30. The purpose was to facilitate liaison with the Singapore Infantry Regiment to whom a Senoi Praaq squadron had been seconded.

31. The Senoi Praaq troopers had in fact been back on operations in the BOA for two months prior to the Labis raid. Confrontation saw the peninsula stripped of conventional troops and GOM of necessity became amenable to various proposals promoting their continued utilization against the MRLA.

32. R.O.D. Noone had a terrible cigarette habit and smoked about three packs a day when not out on operations. And so after putting himself in harms way for a lifetime, he died a banal albeit unpleasant death of cancer. The disease killed him in 1973 just as he was about to settle down and remarry to a very nice Thai lady. He remained Rural and Border Security Officer for SEATO in Thailand until his death. His old friend and comrade in arms Charles H. Ley joined him in Bangkok for his final months as a last gesture of fealty. A meticulous planner to the very end, he is said to have had—in a parting gesture—his ashes spread from an airplane over Chiang Mai province and at a commemorative stone place in Chiang Mai

city's Foreign Cemetery. A memorial service was also held for him in Bangkok at which a eulogy was read by Sunthorn Hongladarim, Secretary General of SEATO. Today his monument is well kept. He was well respected and liked in Thailand and is also remembered in SAS circles.

33. It is interesting indeed to speculate upon whom first asked R.O.D. Noone to assist in the search for Jim Thompson. New light may well be shed upon the mystery of his disappearance in Michael G. Jones' forth coming biography of R.O.D. Noone. This work is expected to examen Noone's involvement in a number of Southeast Asian conflicts and dig up a few skeletons.

34. R.O.D. Noone also made a more discreet visit to neighboring Orang Asli settlements with only his body guard and the aforementioned bomoh, who had been of previous assistance to the Senoi Praaq, in tow. He would not say much publicly about Jim Thompson's disappearance but made it clear that he felt Thompson was not lost in the jungle. Some observers speculate Thompson was most probably abducted by the communists, debriefed and executed. His exact fate remains a mystery to this day. See, Warren, 183, 192-196. Alternatively, a Malaysian source close to the initial investigation suggested to the author that he may well have been killed by people on his own side, i.e. by other Americans. In such a scenario it would have made sense to do the dirty deed outside of Thailand in order to deflect suspicion from the true culprits.

35. There is the outside chance he went to work on one of the many plantations on Borneo and thus disappeared in less than sinister fashion; however, this possibility is thought by former colleagues—none of which have heard from him since—to be slight. His exact fate remains unresolved but he may somehow have overcome his bad reputation and become a Dato only to die later under more peaceful circumstances in a hospital at Kota Kinablu albeit alone and lonely.

36. Not only could one vanish with ample ease in these places but on Borneo body parts are actually considered a form of interior decoration. In a worst possible case scenario Norman Herbolt would thus have met a terrible and somewhat unjust fate. He was, according to some of his less vociferous detractors, not that bad a fellow and had a few redeeming qualities. He had been a flag Lt. to an Admiral in the British Navy and he served R.O.D. Noone well. The man is said to have picked up a bad reputation because he knew how to size people up and was quick to lord over inferiors and kowtow to superiors. He just value maximized incorrectly one time too many and may have preceded in death his aging mother, who was a charming "upper class" lady from Cheltenham.

37. Charles H. Ley assumed R.O.D. Noone's SEATO appointment upon the later's death in 1973. It was an ideal vantage point but R.O.D. Noone was fond of telling friends SEATO did not bankroll sufficient sharp practice. Some say he was actually a bit derisive about the toothless tiger and called it the Sitting Eating and Talking Organization (SEATO).

38. One source mentioned Indonesian planes followed an established flight plan over Johore, suggesting that more than one run may have been made in the Labis area. Thus, once again, we see potential discrepancies between official reports and *post facto* comments made to the author by actual Senoi Praaq.

39. James and Sheil-Small confirm Mackie on this point. They also agree that one plane failed to arrive over the target and disappeared. Its remains have never been found. However, they are not consistent in that James and Sheil-Small report a fourth plane failed to take off due to engine problems. Was it ever repaired? Did it take off later and manage to disgorge its cargo? These questions remain unanswered. See, Mackie, 260; James and Sheil-Small, 148,149.

40. This information is compiled from three separate aerograms from Kuala Lumpur

to Washington, D.C. approved by R.W. Moore, First Secretary American Embassy Kuala Lumpur, dated September 4, 1964, September 11, 1964 and May 28 1965. See Department of State, Central Files, A-165, A-186 and A-936, National Archives, Washington D.C.

41. A Singapore Infantry Regiment patrol had been badly shot up along the banks of a river where it made the mistake of stopping to bathe without posting sentries and so once again the Senoi Praaq was called upon to pull someone else's chestnuts out of the fire.

42. Great Britain is also reported to have considered air and sea attacks upon Indonesian bases. Labis and Pontian were a real sore point. See Mackie 263-264.

PART VII

Senoi Praaq Rehabilitation

In Part VII we complete our study of Senoi Praaq evolution from a handful of troopers into a pair of elite Police Field Force battalions and focus on the unit's present activities. We examine the reconstruction of the Senoi Praaq, analyze its fresh post-Confrontation role in Malaysia's security structure and tie these pivotal events into a new mandate to suppress the revival of MRLA aggression along the Thai/Malaysian border.

We observe Senoi Praaq transition from MIS control into a new affiliation with and absorption by the Royal Malaysian Police and get a glimpse of a new Senoi Praaq commander who maneuvers around endless bureaucratic hurdles to guarantee the unit's survival. Thus, as we arrive at the terminus of this study the Senoi Praaq quest to find a permanent place in the Malaysian order of battle is at last achieved.

Today the Senoi Praaq performs a variety of non-military activities. In a parting glance we assess some of the unit's more mundane contributions to Malaysian contemporary daily life. Our findings indicate it is well within the realm of possibility for bad boys to have a little fun while doing good deeds.

Chapter 21

The New Routine

Indonesia's attempts to isolate Malaysia on the world stage failed after several years of fruitless effort at the United Nations to garner external support for her policy of annexation. Her assaults on the peninsula soon died down due to domestic political tensions emanating from an internal leadership struggle over the Confrontation and other matters.[1] Strong and sustained Malay/British military resistance played no small part in the deterrence of further aggression and it became expedient for the GOI to make peace with the GOM and end the war.

PEACE IS NOT AT HAND

Thus, by August 1966 the Confrontation died away but demand for Senoi Praaq services did not slack off in the slightest. Malaya proper had been depleted of troops during the initial phase of the war and upwards of 27,000 men who would otherwise have been available for duty on the peninsula were deployed to stabilize the situation on Borneo. Subsequent Indonesian assaults on the mainland saw to it that any remaining Malaysian troops were thoroughly committed in the defense of the southern states. In the end this *tour de force* failed from the Indonesian perspective but it had a secondary effect creating a vacuum of sorts which the MCP was able to exploit along the Thai border.

MRLA ACTIVITY DURING THE CONFRONTATION

The Senoi Praaq troopers were indeed stretched very thin for a time and they fought something of a two-front war contending with Indonesian infiltrators in the south and MRLA probes to the north. The MCP did not launch an all out assault but it tested the GOM sorely. Skirmishes between the MRLA and security forces in Perak in tandem with attacks by Indonesian forces in Johore rocked Malaysia. The MRLA for its part tried after a fashion to network with the Partai Komunis Indonesia (PKI). Thus, the MCP had finally managed to secure a limited form of outside sponsorship and it took advantage of this paltry support to resume operations in the Betong Salient area.

Malaysia proper found itself challenged once again by communist subversion. A very real possibility existed that the MRLA would double down, sortie from Southern Thailand where its forces regrouped following their expulsion from Malaysia in 1960 and take back the initiative in the still festering insurgency. Although the MRLA had been beaten in the field and driven out of its deep jungle bases in central and northern Malaya, it had not been destroyed. About 2,000 thousand fighters were reported to have made good their escape to the Betong Salient region.[2] For years afterward a hardcore of men grouped into several large concentrations remained active along the border between the two nations. Conservative figures continued to place MRLA strength in 1964–1965 at about 800 seasoned fighters in addition to numerous support elements.[3]

SANCTUARY ALONG THE MAIN RANGE GIVES WAY TO A BASE IN SOUTHERN THAILAND

In fact the MCP, despite its misfortunes in Malaya during 1948-1960, soon built and operated a *defacto* government in southern Thailand. MRLA fighters took advantage of the wicked terrain, Royal Thai Government (RTG) laxity, a poor system for liaison between GOM and the RTG and the ethnic Malay inhabitant's discontentment with Thai rule. This combination of favorable circumstances and a fervent desire to prevail enabled the MCP to dominate four of the country's southernmost provinces and threaten Malaysian sovereignty well into the 1980s. Narathiwat, Pattani, Yala and Satun provinces were particularly infested by terrorist gangs. Thus, the communist fighters were never fully dislodged by Malaysian security forces or otherwise compromised despite eventual Thai cooperation.

Southern Thailand/northern Malaysia—one of the last wild and wooly areas on earth—was ostensibly under the control of two sovereign, stable and relatively prosperous countries. However, communist party strength in the area was so strong it had the capacity to operate openly, levy taxes and to administer effectively a significant chunk of Thai territory. The MCP managed to consolidate this foothold through its deliberate limitation of MRLA activities on the Thai side of the border. The communists made no effort to confront Thai authorities, they left the minority ethnic Thai population alone and unmolested, and they suppressed banditry in areas under their control. Real "cohabitation" took place here long before it ever did in any French government.

Instead, the MRLA sought out the local Chinese population in the Betong Salient area and, as might be expected, imposed MCP authority. It also went to great lengths to cultivate the ethnic Malay population in adjoining areas and to exploit the fact that these provinces had been Malaya territory until 1919. Communist cadres often converted to Islam, adopted Malay names and married into Malay clans just as they had penetrated Orang Asli communities in Malaysia at the outset of the Emergency.

Consequently, many young people—Chinese and Malay alike—enamored by the communist rhetoric cooperated freely with the MCP. The party's propaganda campaign went so well that one is left to consider that had greater efforts been made along this line earlier in the Emergency the conflict might well have taken a somewhat different course despite the obvious theoretical imponderables that differentiate communism from Islam.[4] Thus, the handmaidens of communist subversion—propaganda and selective assassination *cum* terror—once again served to neutralize resistance and to advance the MCP agenda. Anyone from within these two target population groups—Malay and Chinese—who did not toe the party line was killed or otherwise neutralized.

THE ROYAL THAI GOVERNMENT POSITION

Thai authorities for their part turned a blind eye to these activities, obligingly so for several less than altruistic reasons. The dictates of realism rather than fear of the communists pervaded their thought processes. From a geo-political standpoint a modest MCP presence in the area tended—to the RTG way of thinking—to offset existing radical Muslim influences. Both groups had separatist ambitions but these could be mitigated by playing one group off against the other. Southern Thailand had once been part of Malaya and there was some antagonism between the two governments over this territory. Hence, why not exploit Malaysia's weakness by ignoring the communists.

However, a more pedestrian explanation for this *laissez-faire* arrangement between the RTG and MRLA forces lies in the local Thai authorities' lax attitude towards law enforcement. A significant degree of corruption permeated Royal Thai Police (RTP) personnel who generally were content not to take any undue risks in actions aiming to suppress the communists. There is also some direct evidence to support an assertion that local RTP officials accepted communist bribes and looked the other way for the reason that their leaders profited from the sale of war materiel to the MRLA and MCP.[5] The war literally generated some new types of trade and facilitated commerce that otherwise would not have taken place.

These dealings in war materiel also germinated the seeds of Malaysia's difficult drug problem today which—as we shall see—actually created work for the Senoi Praaq. As the Vietnam war geared up Malaysian Chinese traders colluded with Thai authorities in the procurement of drugs for the many American armed forces personnel who were sent on leave to Malaysia. Some GOM officials finger these Americans as the ones who really provided the impetuous whereby this deadly habit was brought to Malaysia. Prior to this time Malaysia had little drug activity in the country and blatant addiction was rare. To back up their claims they assert the MCP and MRLA always flush with cash derived from the drug trade made repeated efforts to outbid Malaysian government forces in the purchase of scarce provisions from the Orang Asli in the border vicinity.[6]

Thus, the RTG was much more concerned with the communist threats coming from Laos in the north than it was those from Malaysia in the south of Thailand. And so, for Malaysia a tense time ensued as the country essentially remained bracketed between dual external threats—an active one from Thailand and a more latent menace by Indonesia. Fortunately, for the Senoi Praaq another fortuitous set of coincidences had come to pass. Already back in GOM's good graces thanks to its efforts during the Confrontation the unit would soon become the "golden child" of the Malaysian government. Plenty of work was left to be done and management of the CT problem turned into a long-term obligation and a full-time job for the Senoi Praaq.

Chapter 22

The Senoi Praaq in Transition

At this critical post-Merdeka juncture all the pieces fell more or less into place for the Senoi Praaq troopers. They were poised to take advantage of the many attenuating circumstances to surround formation of Malaysia, shed their skin as a private army and metamorphose into a bonafide Police Field Force (PFF) unit. The 1960s were thus a time of great opportunity and they did not hesitate to adapt quickly to the emergent political circumstances and make the necessary deals to survive within the new political order. And so began the tenure of Edmund C.R.Dewsnup as unit commander on April 3, 1963. Dewsnup, an aristocrat by birth and a natural war leader gifted with multiple personal qualities he could summon at will, was equally at home in the world of the soldier or of the diplomat and was someone who lived by his wits.

YOUNG BLOOD RULES

This suave individual, a real killer with a fondness for opera, who came by his talents instinctively, knew how to stay on the right side of the law. He could easily be credited with the jingle "we never killed anyone who didn't deserve to die"—an apt motto for the Senoi Praaq. The troopers of the Senoi Praaq are very proud of the fact they never shot anyone by mistake and their sober prudence in handling their affairs is credited with their ability to avoid alienation of the civilian populations—Malay, Orang Asli, Thai and Chinese alike—in the areas in which they operated. Thus, they never gave the MRLA an opportunity to swell its ranks through appeals to hatred and fear of the Senoi Praaq (as opposed to MCP coercion or the ideological attraction of communist doctrine).

In any event Dewsnup long a force within the Senoi Praaq became the commander who oversaw the transfer of power from colonial and to Malaysian jurisdiction. Dewsnup like Charles H. Ley had had control over the notorious Semai based A squadron and took over generally from Norman Herbolt when Herbolt left to join R.O.D. Noone *et al* at Pleiku.[7] His initial appointment was—by agreement—temporary and for the period of Herbolt's absence. However, under the existent circumstances it did not take long for Dewsnup to convince the powers that be it would not be a good idea to let Herbolt come back to his old job. Skullduggery thus accounted for Herbolt's having gotten the chop and was the reason why he ended up hanging around on Borneo longer than might otherwise have been expected.

Dewsnup was also something of a mischievous rogue—*à la* Jacques Dovey (pseudonym)—who by hook and crook would see to it the Senoi Praaq got back on operations.[8] Much like R.O.D. Noone he really did live for the unit and was addicted to action and adventure. He was also just as capable of intrigue as Norman Herbolt, if not more so, but petty political games—other than those designed to preserve the Senoi Praaq—were not his cup of tea. As an astute politician Dewsnup sought instead to undo the Noone/Herbolt stigma, repair the Senoi Praaq good name and thereby make amends with other police and military forces.

BUILDING BRIDGES

Consequently, a new and very solicitous Senoi Praaq leader in the first few months of his command engaged in shuttle diplomacy and met repeatedly with representatives of military intelligence, the Special Branch[9] and regular police forces. His message "we will cooperate freely with you and keep our noses clean" was a very simple and direct one aimed at trading off further Senoi Praaq pretensions for guaranteed survival. Unlike Noone and Herbolt, Dewsnup was a true realist who actively courted Kuala Lumpur VIPs and of all things sent our old friend Colonel Fargus on a "round trip" junket into the BOA. Needless to say these tactics went over very well, previous hostilities between very real antagonists were soon forgotten and all was eventually forgiven.

In the formal chain of command the Senoi Praaq operated much as before, under the indirect supervision of an extensive oversight mechanism. However, Dewsnup knew if he could just manage to stay on the books, as it were, the Senoi Praaq would retain *defacto* control over the BOA. This was the key to everything. Whether or not one operated independently—and chaperon-free—was a secondary concern. The rest could be worked out later. So, as an integral part of a broad charm offensive he also bit his tongue and made himself agreeable to the secretive MIS. Few people know much about the activities of this entity which had nominal control over the unit ever since 1960 at about the time Norman Herbolt was installed as commanding officer. Captured CTs who did not "renew" themselves, or otherwise make amends to the government, could expect unpleasant dealings with the MIS should they be taken prisoner by government forces.[10]

NOTHING IS EVER EASY

Unfortunately, Senoi Praaq problems extended well beyond British preserves and into the Malaysian side of things. Malay GOM officials were not so forgiving, once pissed off. In this regard, Norman Herbolt had charge of a Malay 2IC named Yusof Khan—a real nasty piece of work—who, while Norman was busy among the Anglos, managed to alienate thoroughly the Malay side of the government.[11] Ordinarily this would not have been a problem but times had changed and by this point the Malays had worked their way into control of quite a few critical operations.[12] Malayanization had been underway for some time and as a result there were, by the mid- to early-1960s, only so many things the British were able to keep amongst themselves. Consequently, preservation of the Senoi Praaq was not simply a matter of smoothing things out among Europeans.

Fortunately, in addition to his professional and diplomatic skills Dewsnup was also very adaptable on a personal level and something of a friendly chameleon. Not only did he manage to soothe the ruffled feathers of Colonel Fargus *et al* but he also adopted local customs and consequently made a very favorable impression upon the Malay leaders of the new government. His extreme pragmatism thus enabled him to blend easily into the changing environment and skirt a pitfall that ended the careers of many British holdovers in the emergent nation state.

THE LAST TUAN

GOM had begun to implement a Malayanization program for her civil and military servants shortly after independence. British personnel on loan to GOM and the Malaysian Armed Forces (MAF) were gradually replaced by local talent as circumstances merited. These individuals knew their time was limited as more and more trained Malaysians would eventually become available to take their places. Thus, quite a few opted to leave in the short term as a forty-year-old would have a much better chance of signing on somewhere else than a fifty-year-old. Their departure created something of a personnel shortage for GOM, one it could ill afford at such a critical moment in the nation's history.

Thus, a convenient loophole was left in place for those expatriates—sufficiently devoted to Malaysia to merit exception—who sought to continue on indefinitely in the service of the GOM. Technically, any "citizen" of the Federation was eligible for employ and thus through conversion to Islam and marriage to a young lady from Penang, Edmund C.R. Dewsnup became a true Malaysian. By dint of this adaptation to local conditions Edmund Colin Ritson Dewsnup metamorphosed into Mohamed Ruslan bin Iskandar Abdullah, the "salve of God," and secured his place as the last Anglo commander of the Senoi Praaq. Like Pat Noone before him he essentially went native, became a prince among warriors and acquired permanent membership in an elite fraternity of great British adventurers.

That he also "lusted" after the CT's did not hurt matters much either. Mohamed Ruslan bin Iskandar Abdullah had been a lieutenant in the British occupation force stationed in the Suez Canal zone of Egypt before signing up for Malaya and he was no cherry. He requested assignment to Malaya in search of adventure and transferred into law enforcement as this was the best way to get into the thick of things and, if personal risk was not a consideration, to acquire real power. Thus, he came up through the ranks of the Royal Malayan Police as a SOVF leader who saw plenty of gritty action prior to joining the Senoi Praaq. He led several SOVF units and had, at various times, been rotated into and out of commands in which his predecessor and successor both were ambushed and killed (along with many of their men) by the MRLA. Their deaths were particularly macabre affairs for these brave men, who after innumerable other unpleasantries, had had their cocks cut off and stuffed into their mouths. In fact, Mohamed Ruslan bin Iskandar Abdullah himself almost underwent this ritual —*se faire passer au coup coup*—one night not long after joining the DOA when his own impetuous behavior nearly got the best of him. So, the stakes were fairly high for the men involved. What a tragedy for the female of the species this would have been as Dewsnup was something of a ladies man who knocked women off their feet with his charm and was a party animal of exceptional qualities.

Chapter 23

Mohamed Ruslan bin Iskandar Abdullah Takes the Senoi Praaq into the Future

Clearly, Mohamed Ruslan bin Iskandar Abdullah was a highly motivated individual with a score to settle and a perfect fit for this sort of job. A deep pathological hatred and persistent psychological need to kill communists engendered by their fiendish ways made him a reasonable man for non-communists to deal with in matters related to the Senoi Praaq. This is an important point since every Senoi Praaq leader's overriding objectives from the time of the unit's inception in early 1955 had been to get and to keep operational and institutional autonomy.

Complete freedom of action in operations and institutional autonomy had been coveted for political and military reasons, not to mention the egos of unit commanders. These twin objectives had—sometimes at the sacrifice of good working relations with the government—been largely achieved. The DOA had its own line in the budget and a vote in parliament through the Orang Asli Senator and it was not hurting for influence. As a consequence the Senoi Praaq command, from a logistics point of view a quasi-autonomous unit operating under the umbrella of the DOA, was not all that dependent upon any other entity. Too bad so many people had their knives out for Senoi Praaq commanders, but partisan politics in Malaysia is among the most vicious in the world.

NEW ALLIANCES ARE MADE

Hence, in a reversal of this longstanding imperative—first pursued so skillfully by Richard Noone and later most awkwardly by Norman Herbolt—in which the Senoi Praaq operated alone, Mohamed Ruslan bin Iskandar Abdullah made it perfectly clear he was prepared to work closely with other organizations and, as appropriate, to farm out Senoi Praaq operational elements. At about the same time he also instituted new policies to broaden the recruitment base of the unit. Heretofore the emphasis had been upon recruiting Orang Asli from the northern region, especially the Senoi Jelmol, i.e., hill people and Senoi Baruh, i.e., people of the plains. Now there were many Jakun type tribesmen from southern regions of the country to be found in the unit not to mention a few Semang as well. Hence, the Senoi Praaq had become not only more reasonable to deal with but also a well-integrated unit that could function efficiently throughout Malaysia—and just in time to participate directly in the later stage of the Indonesian conflict.

Through no coincidence these new qualities Mohamed Ruslan bin Iskandar Abdullah brought to the Senoi Praaq became the unit's quintessential elements in the face of Indonesian incursions directed at the southern states of peninsular Malaysia. Once the war escalated on Borneo, Senoi Praaq detachments were therefore made available for loan to perform recce and other services on behalf of conventional units, and as a result quite a few Malay and British field officers were especially indebted to Mohamed Ruslan bin Iskandar Abdullah for his help when the fighting spread to the mainland. Without Senoi Praaq cooperation and assistance their own forces could not hope to operate with any degree of efficiency—let alone avoid extensive casualties—in the boundless mangrove swamps which enshrouded their operations areas.

Thus began a long sequence of new cooperative ventures between the Senoi Praaq and other armed forces entities and various intelligence operations. Most notable among these was with Special Branch personnel in the establishment of a series of listening posts in those areas near the Thai border still threatened by CTs. The routine was for two Senoi Praaq troopers to accompany a Special Branch officer to a location where Orang Asli were known to live and to remain in the vicinity for a protracted period of time in hopes of developing human intelligence about MRLA/Asal interaction with the locals.

The two Senoi Praaq troopers fraternized with members of neighboring settlements and secured information through guileless displays of friendship not to mention the dispensation of largess. The Special Branch officer was then in a position to process this information and relay it back to headquarters for further evaluation. Ultimately, the entire BOA was blanketed in this manner and thus it became very difficult for CTs to penetrate any depth into Malaysia. Many of their patrols never made it back over the border into Thailand once picked up upon by these listening posts.

This passive form of networked intelligence gathering was also helpful in the long run in growing the organization of the Senoi Praaq. Such cooperation with the Special Branch led eventually to an allocation of a further 80 men in order to form a fourth squadron (to man the listening posts) which made the Senoi Praaq more akin to the usual SAS regiment size. The institutional linkages between the two entities formed during this cooperation would later facilitate Senoi Praaq transformation into a mainstream PFF component. It gave the unit a friendly voice in the administration most notably in the person of Desmond Lawrence—an officer of Portugese/Indian descent who was the Special Branch Director for Pahang State —which would stick with them through the rough times which lay ahead.

SENOI PRAAQ RAIDS INTO THAILAND

Soon after the Confrontation ended on the mainland there was a lull in MRLA activity, the number of CTs thinned out on the ground for a time, and it became necessary to go after them into Thailand. Operations against the MRLA in the Betong Salient region were then extended beyond the borders of Malaysia proper. Senoi Praaq squadrons began to probe deep into southern Thailand in search of those insurgents who had successfully eluded security forces further south. Some raids, actually coordinated from mobile Senoi Praaq command posts set up inside the Thai border, were carried out at the squadron level and penetrated between five and twenty-five miles depending upon the particular Border Agreement in place at the time.[13]

The raids were productive and GOM's money well spent. Very rarely did Senoi Praaq troopers fail to return without a nice bagful of hands to show for their trouble. Heads were also taken on occasion for identification purposes as were the occasional pair of ears as proof of a kill.[14] Hands though were of greatest value. If a match for the finger prints could be found this would allow the Special Branch to go after the families of the individual in question.[15]

The routine was fairly straight forward but these operations were not always easy to carry out. There was a lot of legitimate activity in the immediate border area and it was often difficult for the Senoi Praaq troopers to differentiate tracks and traces left by the MRLA from those of woodcutters, traders and others who came and went on lawful business. By the same token, if their own activities could be picked up and become common knowledge among the locals, the MRLA fighters would not take long to register their presence. So, the drill was for the Senoi Praaq troopers to pick an operations area in which they were free to move about with some degree of security and thus to transit quickly the border region.

Once established someplace fairly remote they would take to the high ground, pick a good sized tree and therein build an observation platform from which to place the area under surveillance. Dusk was a good time to pick out wisps of smoke from cooking fires which in isolated and uninhabited regions of the Betong Salient were almost sure to emanate from MRLA camps. Once positively identified as such the rest goes without saying.[16]

Mohamed Ruslan bin Iskandar Abdullah himself led more than one of these missions and made every effort to eliminate the threat posed by MRLA remnants. However, the Senoi Praaq could never quite manage to close out the communist operation on the Thai side of the border. Its roots ran too deep for limited raids to do more than relieve the outward symptoms of the malignancy. And further, by the terms of the border agreement the Senoi Praaq could not go after the enemy in urban areas where the higher leadership element was to be found. Nor were large-scale joint Malaysian/Thai operations—negotiated as a sequitur to a general border arrangement—later in the 1970s able to flush out the communists or force an MCP surrender.

The MCP was too entrenched at this point to be rousted by a modicum of brute force or used up through low intensity attrition. Eradication of the communists required either a comprehensive political solution to the problem or the commitment of a surreal conventional force *à la* Vietnam. Unfortunately, the necessary political will let alone materiel resources—to take matters to such extremes—did not exist for either scenario to be considered seriously.

GOM and the RTG did eventually put agreements into place sanctioning cross border missions of some substance but the Emergency would fester none-the-less for many years to come. A stalemate developed in which regular army units were posted along the border at strong points in a static defensive role with the Senoi Praaq and other police field forces positioned behind them in a sweep mode.[17] As always, some of the enemy continued to get through from time to time and there were a sufficient number of incidents to guarantee members of the Senoi Praaq steady employment.

SENOI PRAAQ INCORPORATION INTO THE ROYAL MALAYSIA POLICE

Malaysia and the sword of Damocles would never become strangers to one another and so there could never be any question of a stand-down. Unilateral decommission of her armed forces was not an option despite the hazards involved with too great a militarization. This posed a potential problem as the country was basically armed to the teeth and most of its finely honed weapons were apt to cut both ways if given a chance to do so. A balance had to be struck in Malaysia between military strength and political stability. Hence, in the interest of long-term regime stability steps were taken to homogenize the civilian side of her defense apparatus and forge a more streamlined organizational structure.

By implication Malaysia's order of battle had no permanent place for a private army as such yet the Senoi Praaq could not be disbanded. The solution to this dilemma was to gather the Senoi Praaq and other para-military units like the Borneo Border Scouts and VAT 69 into a unified command. This combination would preserve the Senoi Praaq in function but not form, and it was the approach the government chose to resolve a most sensitive issue—the warlord syndrome. Indeed, R.O.D.Noone, Norman A. Herbolt and Mohamed Ruslan bin Iskandar Abdullah were lords of the BOA with the power of life and death over anyone with whom they came into contact.

Thus, it was with some degree of discretion that the Senoi Praaq, on February 1, 1968 after eight productive years in which little about the unit changed, was transferred albeit very gingerly from MIS to nominal RMP control.[18] At this point Mohamed Ruslan bin Iskandar Abdullah negotiated and put into place the protocol whereby the Senoi Praaq eventually would fully amalgamate into the RMP. He then relinquished command of the unit to Charles Lim and moved back into the DOA from which he would operate until September 1980. Commanders and men alike upon retirement from the Senoi Praaq were guaranteed employment by the DOA if they chose to remain active professionally. Many took this option.

There was, however, one drawback to these administrative schemes which did in fact have a negative effect on the overall Orang Asli power position. The Senoi Praaq already detached from the DOA saw its separation from its parent organization turn into a divorce like situation. This continued martial engineering was a necessary step, everyone recognized it as such, but this final manoeuver did permanently fracture the formal Senoi Praaq/DOA connection. The Senoi Praaq was irrevocably separated from the DOA which had begun some time earlier its own series of moves into and out of Home Ministry control and towards a lessor bureaucratic orbit.[19]

EMPIRE FADES AS CHANGES ARE FINALIZED

Control over the Senoi Praaq thus passed forever out of British hands and for the moment into the grasp of the loyalist Chinese mafia which ran the Special Branch. Extensive behind the scenes activity commenced at this juncture as various groups maneuvered for power. A string of Chinese, Malay and Indian commanders came and went—Charles Lim, Pritam Singh Gill, Othman Haji Dahwan and Chiam Ah Poi are they key names. None lasted very long and a further period of adjustment thus ensued in which once again little about the unit changed from an operational standpoint.

In the meantime the Senoi Praaq continued a series of spatial moves which had taken it from Gombak to Kuala Lumpur by way of the old SAS barracks at Wardieburn, Setapak and at Camp Lam Jung on Jalan Maxwell in Kuala Lumpur. Next the unit was stationed at Kroh in Perak where it was placed technically under the administration of the North Brigade Police Field Force in Ulu Kinta, Perak; however, it was not "integrated" yet into the RMP itself. Only in 1974 did the GOM see fit to initiate any fundamental change absorbing the organization into the corpus maximums. Syed Zainal Abideen Alsagoff's command—which began on September 1, 1975—marked the moment when the Senoi Praaq began to stabilize and experience a steady flow of Malay commanders.

This arrangement involved a trade off of sorts in which the RMP effected the final takeover of the Senoi Praaq through acquisition of its paper. The Senoi Praaq troopers were obliged to exchange various unit insignia and their regimental colors for those of the RMP in return for the buyout of their contracts (at the full RMP rate).[20] They adopted RMP ranks and unit configuration and became Battalion 19 of the PFF. However, as a concession the men were allowed to retain their distinctive red berets which the SAS had given them the right to wear at the time of their formation (prior to the SAS departure from Malaya). Things went so well that three years later the Senoi Praaq—upon completion of its final probationary period—was expanded into two full strength battalions.

September 4, 1977 saw the Senoi Praaq parlay their RMP foothold into a major role in Malaysia's internal security and law enforcement structure. Battalion 19 acquired a twin through levy of a second detachment of Senoi Praaq which, upon formation, became Battalion 20 and took control over the old British army base some distance south of Keroh at Bidor, Perak. In this expansion the Senoi Praaq saw its force strength increase from 300 or so to a 1,000 plus men, but the new arrangement rendered the troops concerned easier to control through homogenization and the creation of a dual command. Now there were two commanders who could be played off one against the other if need be by the political leadership as opposed to one all powerful figure among the Orang Asli. All in all this very clever arrangement satisfied most everyone concerned. The expansion through consolidation created opportunities for system-wide promotion and a much bigger budget.

Chapter 24

The Senoi Praaq Today

Phased incorporation into the RMP marked a new beginning for a reformulated Senoi Praaq and these were good times generally for the Orang Asli people. Yet, Malaysia herself remained under a lot of stress and the late 1960s were difficult for most other Malaysians. As the situation in South Vietnam went from bad to worse GOM found itself in a cold sweat over the American setbacks. CTs had already been observed to be in the possession of American type weapons believed to have been captured in Vietnam and smuggled south through Thailand for use in Malaysia.[21] If the communists could prevail over a superpower then what chance did Malaysia have if the domino theory turned out to be true?[22]

THE WICKED NEVER REST

Fortunately the Senoi Praaq were in high spirits and ready for action when the Emergency picked up again in the late 1960s in what became a pattern of on and off again communist raids emanating from Thailand. 1969 saw, subsequent to the Communist Party of Malaysia's (CPMAL) 1968 directive calling for a renewed general offensive, a marked upsurge in MRLA activity and is billed as a vintage year for the MCP in which it made yet another concerted attempt to the overthrow of the GOM. A revamped MRLA trained in new Viet Cong tactics emerged from the Betong Salient to establish bases in the northern states and take the fight to Malaysian soil once more. These areas subsequent to communist re-infestation had to be purged anew of the red menace and, consequently, for some years to come large tracts in north peninsular Malaysia persisted as "black areas," or exclusive military zones.[23]

There were other flare-ups in the years to follow and some raids were successful in the sense that the guerillas actually hit a variety of important targets. They managed to sever repeatedly the east-west highway in north Malaysia, destroy numerous semi-tractor-trailer trucks and sabotage railway lines around Penang. Military installations per se were also struck with some frequency and a number of police posts sacked. The Central Polis Pasukan Hutan (PPH) Brigade Headquarters at Jalan Pekeliling in Kuala Lumpur and the Royal Malaysian Air Force strip at Sungai Besi were targeted. A number of weapons were lost in these engagements, a few of which were bold and brazen.

The toll extracted in human life was substantial as well and the GOM took quite a few casualties. Several high-profile security force officers were also assassinated. MRLA hit squads thus pulled off several slick coups and made Perak, Kelantan and Pahang dangerous territory even for the most senior officers. The third IGP and the senior police commander of Perak state were both assassinated.[24]

Malaysia and her people continued to suffer much from these depredations. One of the last big spurts of MRLA activity came about in 1974-1975 as the Vietnam situation looked lost. MRLA patrols lurked in the vicinity of the many jungle forts speckled throughout the BOA. Frequent contacts with the MRLA were reported by the garrisons of these posts. The MRLA once again actively targeted the Orang Asli. In one memorable raid the entire population of an Orang Asli village living near Fort Gob in Kelantan state was abducted, driven 50 miles, and spirited away into Thailand.

THE SECURITY SITUATION STABILIZES

Only gradually did security forces manage to eliminate the communist terrorists. Persistent Senoi Praaq patrolling in conjunction with a key infrastructure project—the Temenggor dam—finally snuffed out most MRLA maneuvers any distance from the immediate Malaysian/Thailand border. The Temenggor River had long been a favorite infiltration route which the GOM forces had been unable to seal off totally. Not only did the dam and its flood waters block guerilla passage but these impediments also created a static target which attracted CTs like a magnet. As a result, many who had expected to carry out a sabotage raid against the facility were easily hunted down and killed.

There would be more tense times—particularly after the collapse of South Vietnam—but ultimately it became apparent the domino theory though an inventive concept was a phenomenon that would not play out beyond Cambodia and Laos. In Malaysia the communists began to loose their will to fight. The MCP made the transition into the CPMAL but later it fell prey to factionalism. Discord among its leaders weakened the organization beyond the point from which it could ever expect to recover. Global communism itself was destined to fall out of vogue and at last many of the party's old Malaya hands would see the need to settle with GOM.[25]

Malaysia went on to become an economic success story and very few Chinese let alone Malays were willing to carry the torch. So, the Emergency just withered away and died on the vine so to speak. Senoi Praaq troopers still active in the 1980s report they virtually ran out of opportunities and were put in the awkward position of having to let one or two CTs on patrol escape from time to time just to be sure their movement did not die out totally. The Senoi Praaq over the years had turned into a fantastic "cash cow" for the general Orang Asli population and it made no sense to kill the golden goose through too much bloodlust. Only in December 1989 did GOM and the MCP finally sign a peace treaty at Haadyai in southern Thailand, closing the door on the communist insurgency in Malaysia.[26]

THE SENOI PRAAQ ROLE EVOLVES

As it turns out the Senoi Praaq troopers needn't have worried about their jobs. Malaysia for all of her good fortune and natural wealth was beset by any number of problems—drugs, countraband in general, illegal immigration and natural disasters are only a few which required the attention of a large police force. The ususal national security issues associated with Malaysia's high politics referred to heretofore in Part I hardly had time to enter into the picture.

Hence, with the termination of the Emergency the Senoi Praaq reoriented its traditional counterinsurgency role and assumed some non-military responsibilities. To be sure its troopers still carried out operations in the BOA and prowled around the border area with Thailand but they had any number of other opportunities to put their special skills to use. Consequently, they reverted to a variation of the pattern first established by the last Tuan and began to cooperate with other entities and in this case organizations outside the military and paramilitary communities.

Malaysia is particularly indebted to the Senoi Praaq troopers for the deterrent effect they have on timber poachers now that the jungle has been cleared of communist terrorists and made safe for ordinary thieves. Their interest in conservation is self-serving, however, given that the damage associated with illegal logging activities is also quite harmful to their own tribal communities. Defoliation creates conditions for flash floods—of which there have been a number in recent years during monsoon season—bringing horrendous torrents of water crashing through Orang Asli communities with tragic loss of life as well as disruption of customary business and social activities.

The Senoi Praaq role in these matters is not entirely preventive, either. Once a village is swept away its troopers are expected in their search for bodies to bring to bear those same supernatural powers which protect them on the battlefield. Victims of these tragedies more often than not are buried by mud slides and all traces of scent have been washed away. Tracker dogs cannot find them under such circumstances. Resort to the trooper's black magic is just about the only means to effect a recovery and the authorities do not ask any questions.

And so, even during times of peace the Senoi Praaq continue to trade in death, death of the innocent and the not so innocent. Malaysian law is draconian and rule enforcement in the country no joke as many offences are capital affairs. Drugs are the classic example—for which Malaysia has been subject to much unfair criticism —and for the Senoi Praaq apprehension of a trafficker is just as good as a kill. Woe be unto its customers—revolutionaries, criminals and ordinary citizens alike; one's particulars are of no concern to the Senoi Praaq troopers, Malaysia's killer elite.

In the last analysis if you come into contact with them on business you're either already dead or you're most probably gonna die, soon.

Part VII: Notes

1. The Indonesian military crushed a coup attempt by the Partai Komunis Indonesia (PKI) and sharply curtailed President Sukarno's power position. Together, these events helped to take the steam out of Indonesia's drive to subvert Malaysia. See Mackie, 308-324.

2. Noone, 198.

3. Given the problems associated with arriving at an accurate tally no one knew the exact number of enemy fighters—only that their presence was very real.

4. Conventional wisdom dictates the Malays would never have joined in any numbers due to Malay apathy towards the Chinese, apathy towards a godless belief system, etc. Yet, this line of thought does not explain at all why the MCP was able to make inroads during later years among the Muslim Malay majority in southern Thailand. Thus, the issue of compatibility between Islam and communism is not cut and dried and remains open to debate. Fortunately, MCP leaders never really trusted those few Muslim Malay converts who readily joined the MCP in the early years. Hence, they did not allow Rashid Myden and others an opportunity to make real inroads into the Malay population-at-large. Had they been more open to such a course of action near the beginning of the conflict who knows what might have happened. A more favorable disposition on the part of Muslims in general towards the communists may well have tipped the balance in their favor.

5. In one particular instance a Thai police chief protected a business run by his wife that supplied uniforms to the MRLA.

6. For that matter the real problem lay in the absence of open access to war materiel rather than money *per se*. Money was not the obstacle. Be this as it may, one simply did not derive this quantity of funds through extortion/taxes alone.

7. The Semai are also the stoic Orang Asli tribe described by anthropologist Robert Knox Dentan as the quintessential non-violent people of Malaysia. It is thus ironic the Semai from A squadron turned out to be the most efficient killers ever to serve in the Senoi Praaq.

8. Dovey (pseudonym) and another police officer named Berick (pseudonym) were well- known in Malaya. The pair sometimes played catch with explosives for kicks when not out on operations. One night they were both drunk and fumbled a grenade which Berick (pseudonym) just managed to toss out a window before it ignited. Somehow the incident was passed off as an aborted CT hit and Berick (pseudonym), of all people, acquired a medal under dubious circumstances. The irony of the story is they both—Dovey (pseudonym) more so than Berick (pseudonym)—deserved awards for bravery but not for that particular action.

9. The Special Branch is that division of the RMP responsible for the collection of security intelligence as it pertains to internal and external matters. As such it monopolizes the intelligence function for the GOM.

10. Surrendered enemy personnel (SEP) were shown a lot more courtesy in this regard. The MIS was very short lived, met a quick end through fratricide and is now defunct. "Renew" is a more apt euphemism than the American adaptation "Chieu Hoi" or Open Arms for changing sides. Perhaps that's why it worked so much better in Malaya than Vietnam.

11. There were actually two Yusof Khan(s) and they are not to be confused. The "bad" Yusof Khan mentioned above was a sinister type who did not have much real power outside the influence he derived from his Senoi Praaq affiliation. This despicable individual was Norman's malevolent alter ego and an evil twin of sorts who made Herbolt look like the lesser of the two evils. Between the two of them and their dirty deals they very nearly managed to wreck the Senoi Praaq. On the other hand, the "good" Yusof Khan, a well-respected paramilitary figure who controlled the Police Field Force based in Ipoh, had a fine rapport with the Senoi Praaq. This Yusof Khan played no small part in helping Dewsnup to get the Senoi Praaq back into operations and he is not to be confused with Norman's co-conspirator.

12. Unlike the French who upon decolonization merely moved into the back offices, the British actually gave up immediate control of their colonies—albeit very slowly. Their power migrated to a much higher, global level of authority and they found as far as Malaysia is concerned a symbiotic repository in the Commonwealth.

13. The Orang Asli have a very elastic concept of time and one supposes the same is true for space as well. In that members of the Senoi Praaq are prone to disappear for long periods it is only reasonable to suppose that patrols advanced as far into Thailand as necessary to get the communists. Be this as it may, the Thais over time made an effort to restrict the depth of penetration into their country and the legal limit dropped from twenty-five to five miles. See Chandran Jeshurun, *Malaysian Defense Policy: A Study in Parliamentary Attitudes 1963-1973* (Kuala Lumpur: Penerbit Universiti Malaya, 1980), 86.

14. Here it is expedient to note communist Chinese defectors who went over to the British side and Borneo tribal peoples exhibited by far the greater preoccupation with the practice of taking heads—the Borneo tribal peoples ostensibly in keeping with traditional cultural practices and MRLA defectors to help guarantee a smooth self-renewal process once they found themselves at the mercy of security forces. Typically a lot of reward money was at stake and it was better to be safe than sorry when making claims about having turned on their comrades.

15. The Special Branch kept meticulous files on known or suspected CTs and did not hesitate to put pressure on those family members it managed to get a hold of in Malaysian urban areas. That—the follow up—is why identification was so important. In return the MRLA operatives were particularly keen to get their hands on Special Branch personnel and offered handsome rewards to anyone who could deliver a specimen.

16. However, this tactic as practiced by the Senoi Praaq should be distinguished from the reckless execution of interdiction fire favored by South Vietnamese and American forces in Vietnam.

17. The SAS also moved back onto the peninsula and set up shop just south of the Thai border for a time in an effort to lend support to these endeavors. Ostensibly on a "training mission" the SAS forces as usual were engaged in the most realistic form of instruction. See Cross, *A Face*, 64.

18. The RMP thus took over immediate operational control and would gradually assume direct administrative control as well.

19. The DOA *cum* JHEOA has had various homes and been bounced around over the years from a variety of ministries. They include the Ministry of Education, the Ministry of Land and Mines, the Ministry of National and Rural Development and finally the Ministry of National Unity and Social Development. In early 2001 JHEOA once again was put under the umbrella of the Ministry of Rural Development.

20. Full rate pay was achieved several years later but in stages.

21. See Cross, *A Face*, 120.

22. Tunku Abdul Rahman in particular was quite worried about the effect an American defeat would have on Malaysia and confided his fear some years earlier to Ambassador Nolting while on a visit to Saigon. He is reported to have said at dinner one night: "If you fail here we will have to start all over again in Malaya." See Frederick Nolting, *From Trust to Tragedy: the Political Memoirs of Frederick Nolting, Kennedy's Ambassador to Diem's Vietnam* (New York: Praeger, 1988), 37, 38.

23. The absence of any notable material improvement in the security situation on either side of the Malaysian/Thai border was the subject of much concern at the time and subjected to substantial military analysis. Kedah and Perak were particularly threatened by a resurgence of MRLA activity. See Jeshurun, 89-90.

24. Between 1970 and 1975 twenty-eight police personnel were assassinated. Tan Sri Abdul Rahman bin Hashim (killed in K.L.) and Tan Sri Koo Chong Kong (killed in Ipoh) respectively were the two most senior victims. The Second Emergency was not brought under control until about 1976 with the breakup of numerous CPMAL cells and the detention of several thousand subversives.

25. Chen Peng never did give up his puritanical ideas about communism and he remains an unreconstructed communist. However, the majority prevailed and terms were sought. Chen Peng is reported recently to have been in Australia having his brains picked by academics. So, there may be some interesting literature in print in the not too distant future which sheds new light on the MCP/MRLA account of the Emergency. In the interim Michael G. Jones has an interesting article out about lunch conversation & shop talk with this individual. See Mike Jones, "Mangoes and Chen Peng," *Australian Veterans Association (Malaya and Borneo) Newsletter*, August 1998, 20-21.

26. This agreement finally came about following the withdrawal of what little Peoples Republic of China (PRC) support the MRLA had managed to acquire over the years. As stated the MRLA never did benefit from much PRC assistance even though Chen Peng was a Maoist. Logistics, the tyranny of distance, and perhaps China's reluctance to expand prematurely her sphere of influence within Southeast Asia are just a few of many reasons why the MRLA went begging. *N'import*, there were definite limits to his support from the PRC even before Mao's death and the belated collapse some years later of the Union of Soviet Socialist Republics (USSR) all but finalized the MRLA's fate. Communism—hardcore communism in particular—was on the wane and the MRLA didn't stand a chance once GOM/PRC diplomatic relations stabilized in the post USSR era.

Conclusion

Death Waits in the "Dark": The Senoi Praaq, Malaysia's Killer Elite examines a unique jungle fighting force levied from the Orang Asli tribes of peninsular Malaysia. The specifics of its mission—a closely guarded state secret—center upon the preservation of Malaysian national security to the exclusion of all other considerations. However, as the Senoi Praaq is already a virtual legend in Malaysia this research does not, despite its novelty, expose a secretive organization. Instead, it provides a highly sensitive operation with its first in depth exposure to Malaysians and non-Malaysians alike.

Our analysis thus takes readers through the birth, evolution and development of a small experimental operation that grew into a private army under British control and today is post-independence Malaysia's most valuable paramilitary apparatus. Presently, the Senoi Praaq operates under RMP auspices as a member of the GOF. It accounts for two of seventeen battalions organized into five brigades—Northern, Central, Southern, Sabah and Sarawak.

The exciting and sometimes gruesome ground level view of the Senoi Praaq's many exploits provided directly to the author by veteran fighters is interlaced with information about policy and more than a few backroom deals extracted from archival sources in Malaysia, Great Britain and the United States. These documents, a good number of which have been declassified recently, offer an explanation for the rationale behind extension of the Senoi Praaq license to kill that at one time was valid all over Southeast Asia. They also enable one to extract several poignant lessons from Senoi Praaq accomplishments when these are considered within a broader international relations context and beyond the realm of mere jungle fighting.

Hence, we see through Senoi Praaq eyes a multitude of zesty events reflected in a musty paper trail which together form an odyssey of epic proportions nearly lost by formal history to popular myth. The adventure of the Senoi Praaq troopers begins during the Malayan Emergency in the forbidding jungles of Malaysia's main range, works its way up to the sortie hosted by the United States in the central highlands of Vietnam, onwards to the islands of Borneo and Singapore once the Indonesian Confrontation is in full swing, and ultimately back to Malaysia proper.

However, inasmuch as the wicked are not the only ones who never rest, Senoi Praaq exploits, as extensive as they already are, do not end with a mere return to full circle. No, that would be too easy by far. The original Senoi Praaq foe, the MRLA, though beaten in Malaya, retained organizational integrity and retreated into southern Thailand where it would rear its ugly head repeatedly for the next twenty years. Of necessity, the Senoi Praaq then settles into a long routine of skirmishes with and cross border raids against residual communist forces based in a remote jungle area honeycombed with MRLA bases.

This holding action, a veritable war in and of itself, is perhaps the least known low intensity campaign of the twentieth century. In scale it lies somewhere between the high profile clashes of southern Lebanon and the more limited probing reported between antagonists on the Korean peninsula. The tyranny of distance, skillful limitation of media and other types of publicity in conjunction with constant Senoi Praaq pressure combined to see that the communists were kept out of sight, out of mind and eventually put out of action.

The study tracks the Senoi Praaq rise to prominence through three general phases of action in three different settings across a time period of nearly a half century. First, the struggle by Great Britain and her Malay/Orang Asli allies during the Malayan Emergency against the MCP is brought into focus. The rationale for Senoi Praaq formation is then examined in some detail as is the course of the fighting which took on a life of its own and had to be very carefully snuffed out in the early 1960s lest it continue to feed upon itself. In this initial phase the nascent unit begins to experience some teething problems, not to mention sibling rivalry with other security force units, which see its top leadership more or less banished from the peninsula much as a naughty child might be sent off to the woodshed. Like a bad dream some members of the unit are destined to take part in the sordid American debacle in Vietnam where lessons from the Malayan Emergency were to be tragically misapplied.

Then, during a second and very unpleasant phase in the Senoi Praaq life cycle the intrigue continues but not among various British and Malay security forces but rather between British, Malay, and Orang Asli operatives and their new American friends who have just managed to hasten the French exit from Indochina. This feat was really quite an accomplishment given that Indochina was a virtual French overseas province so to speak and someplace the French did not simply walkaway from without a real struggle. That the United States could supply the French so freely in Vietnam while they themselves were tied down during the Korean conflict only to turn on such erstwhile allies (the sponsors of their very own independence) once the decks were clear was truly a fine example of power politics in play.

The vicious high politics associated with BRIAM, the regime of President Ngo Dinh Diem and the many tentacles of the United States Leviathan thus come to light much like dirty laundry which is finally aired. The United States turns out to have had no real friends, nor any policies, in Vietnam—only interests. Thus, we see how the Americans cannot resist competing with BRIAM over Vietnamese attentions and how the fallout from this screwing around provokes the Senoi Praaq decision to withdraw from Vietnam and move on to more productive pursuits.

To encapsulate, the overall military initiative within which the Senoi Praaq troopers seek to ply their trade rapidly deteriorates into a primitive "go-here-kill" kind of situation. Pleiku province, the Senoi Praaq tactical operations area, becomes particularly unstable. Smart money stands down rather than risk consumption by the mayhem but this does not mean the unit is out of a job.[1]

All Southeast Asia is unstable by this point and there is no shortage of opportunities for the Senoi Praaq. Great Britain and Malaya are at war with Indonesia on their own home ground over territories on the island of Borneo and formation of the Malaysian Federation. Senoi Praaq elements in Vietnam thus disengage not long after the assassination of President Diem and subsequent neutralization of BRIAM influence and with R.O.D. Noone at their head make their way to Borneo.

The British will, after a fashion, find satisfaction on Borneo for their shoddy treatment by the Americans in Vietnam—to whit, a subsequent and successful re-application of those very same methods developed in Malaya and proposed for naught by British, Malay and Orang Asli operatives in Vietnam. This time around though the good guys were pitted against a sovereign foe with much deeper pockets than the MCP. The proof in the pudding, however, was their choice of tactics and methods rather than abundance of support.

Indonesian aggression against Sabah, Sarawak and the Sultanate of Brunei is arrested with ease and Malaysian statehood guaranteed without inducing Vietnam style escalation. And so, well after the American fiasco had begun to unfold in Vietnam, the Indonesians striking from the Kalimantan side of Borneo were thwarted by proven techniques which won the day anew just as they had done some years earlier on peninsular Malaya. The Senoi Praaq, in this engagement assume control over a mixed group of Borneo bred tribal fighters and these troops are then used very effectively against the Indonesians. Thus, the Indonesian confrontation demonstrates only too well that suppression of the communists in Malaya was no lucky accident.

True, Malaya, Vietnam and Borneo are different places but arguments about the similarities and dissimilarities among these environments really are beside the point. The operative principle in irregular warfare in general, and guerilla war in particular, holds that the methods to be applied must be synchronized with the resources at hand and desired ends. It does not take mental acrobatics to grasp that the inverse of this dictum is equally untrue as regards the suppression of most low intensity conflicts. Awkward and over zealous efforts by United States policy makers to suffocate the communists in Vietnam merely ended up fanning the flames of their resistance.

The Senoi Praaq (and other Malay/British troops) for their part were so successful on Borneo at not reverse engineering this axiom that together they drove the Indonesians, operating from the built up regions of Sumatra and Java, to strike a series of desperate blows directly at the peninsula itself. This dramatic turn of events coincides with their own re-employ on the mainland, reinforces the decision to rehabilitate politically the Senoi Praaq and guarantees pardon by the GOM for all previous sins. In a paradox of sorts the escalation turns out to be good for business in Malaysia unlike the situation in Vietnam.

This new assignment takes the Senoi Praaq troopers through a complex series of joint operations with conventional forces on the peninsula's east coast, inclusive of central Johore state and Singapore Island. It also paves the way for full scale operations to resume in the north of Malaysia which sends them numerous times into southern Thailand. Many of these raids were lawful only in the classic sense the strong do as they wish and the weak suffer as they must. As ever the elusive MRLA is their primary prey and remains so for another twenty plus years as the Emergency did not actually end until 1989.

From the late 1960s onwards the Senoi Praaq thus experiences a third developmental phase and prospers accordingly. Its composition changes to reflect a broader mix of west Malaysian tribal peoples and its budget rises. The unit also slips its reputation as a private army, acquires a prestigious new police affiliation and sees its personnel expanded twofold. These developments impose a contemporary face on the Senoi Praaq, usher the organization into the 1990s and take Malaysia safely into the second millenium.

Therefore, the Senoi Praaq today stands at peek readiness and is better equipped than ever before to play the counterinsurgency game. Of course, multi-state sanction for such activities by the Senoi Praaq no longer exists and its operations are limited to the territory of the Federation of Malaysia. It is a peaceful time in the region apart from the high profile disturbances in neighboring Indonesia which are related to separatism and affect Malaysia little, apart from the problem of illegal immigration. Malaysia is, for all intents and purposes, a very safe place to visit in spite of or perhaps because of her high degree of military readiness. Naturally, one need not be afraid to make a trip to the land of the Keris for the Senoi Praaq are kept on a tight leash and are not interested in casual visitors: just don't bore them too much.

However, if violence *per se* is sometimes said to ever be on the lookout for legitimacy then in this particular instance the Senoi Praaq troopers are not at much risk for developing myopia from excessive eye strain. There is still plenty of work for them to do in the form of routine civilian police activity while they bide their time waiting for the next campaign in which to make use of their special talents. Illegal immigration, illicit drugs, unlawful logging activity and the usual banditry along the Thai border are not omnipresent but they do provide a steady source of employment for Senoi Praaq troopers in fairly remote areas off the beaten track.

In this respect the basic geography of Malaysia's remote hinterlands is intact despite Malaysia's steady progression of economic development projects. The BOA for that matter embraces some of the most tortuous jungle terrain on earth (riddled with disease and parasites) in which a healthy unencumbered individual often needs hours on end to travel a single mile. Life expectancy in the jungle is very short for all but the most hardy and few people are adapted to function naturally in this treacherous environment.

Similarly, Malaysia herself although at peace today lies in an environmentally cruel part of the world that has not changed much since the days of Parameswara. History makes it only too plain for us to see that there is no future for the weak or naive in this pitiless place. That so many people can do profitable business or buy a vacation package and have a good time in Malaysia is misleading and something of an anomaly made possible only by the GOM's resolve to resist the twin evils of internal subversion and encroachment by foreign powers. Malaysia through the projection of strength has no outlandish quarrels with her immediate neighbors, accepts the rule of law and embraces democratic practice for the sake of her own stability just as she welcomes overall United States hegemony in the region but not at the price of becoming a lowly pawn.

Hence, broader geopolitical circumstances and genetic qualities inherent in the Orang Asli, which make them resistant to malaria and other maladies that would quickly kill off outsiders, dictate the contemporary Malaysian state entertain an ongoing military relationship with her indigenous minority peoples. It took centuries for the Orang Asli to adapt physically to their jungle environment and centuries more for the right set of political circumstances to come together in which an organization such as the Senoi Praaq might be created and nurtured into the perfect jungle fighting force. Herein lies the rationale for this *modus vivendi* in which political fortitude is married to willing and uniquely able servants of the state.

GOM cannot duplicate the Senoi Praaq by recruiting another group of Malaysians trained in jungle craft and injected with prophylactic medicines. Without doubt geography will remain the friend of the Orang Asli for many years to come and will continue, as long as the world is plagued by realism, to guarantee them a distinctive role in Malaysian life. Thus, anyone undertaking an intelligence calculation involving Malaysia best be aware that whoever controls the Orang Asli controls the deep jungle and to control the deep jungle is to control much of peninsular Malaysia.[2]

As players in the game of power politics the Orang Asli and the Senoi Praaq warriors are, in the final analysis, a neglected albeit critical factor in the analysis of Malaysian national security issues. True, the new and revamped Senoi Praaq is a kinder gentler paramilitary organization tutored in the niceties of law and in civilian police enforcement techniques but old habits are so very hard to break. Even today despite a reorientation of its primary mission death waits in the "dark" deep Malaysian jungles ever so patiently for all those who would dare to try and beat Malaysia's killer elite at their own game.

Conclusion: Notes

1. That R.O.D. Noone, the Senoi Praaq commander, really was a humanist who cared deeply about the Orang Asli and his own men in particular is well illustrated by his refusal to allow any trooper whose siblings had been killed off during the Emergency to accompany him to Vietnam. Those individuals who were the only living male member of age in their own family group were left behind in Malaya and not taken to Vietnam. At the time quite a few Senoi Praaq were very unhappy about this precaution but even now, forty years later, they can still feel his love for them and obviously Noone's commitment lingers on in their minds. His deep sense of responsibility is also the underlying reason why he welcomed the British decision to disengage from Vietnam. He knew deep down in his heart that not only was the situation hopeless but that all of his men would surely be killed if they stayed to fight the good fight.

2. The Malayan peninsula including Singapore covers an area in excess of 53,000 square mile of which about 30,000 are given over to jungle.

Glossary

Aboriginal Peoples Ordinance. Enacted in 1954, this ordinance (often referred to as Act 134) is designed to safeguard Orang Asli interests. However, of late there has been some criticism of its provisions as being overly protective.

Asal Organization. The Orang Asli based MCP/MRLA support organization, sometimes referred to as Asal Clubs.

Asli. Derived from Asali meaning original. The word has Arabic roots.

Barisan Nasional Front. The ruling coalition group dominated by UMNO, the MCA, and the MIC. UMNO, MCA, and MIC stand for the Malay, Chinese, and Indian based political parties respectively. Politics in Malaysia is ethnicity based. The Barisan Nasional Front often referred to as the BN has ruled since 1957.

Belukar. Secondary jungle (extremely congested).

Bomoh. The traditional medicine man cum sorcerer who plays a crucial role in the lives of both Orang Asli and Malay.

Bumiputra. This term translates as "sons of the soil" and is employed by the Malaysian government. It comprises Malay and Malay-like peoples inclusive of the Orang Asli.

Can Bo. Viet Minh political cadres.

Cao Dai. A Vietnamese religious sect that ascribed to a mixture of Buddhism, animism and Catholic hierarchy founded in 1926. The locus of Cao Dai power lay in Tay Ninh – A South Vietnamese province and provincial capital.

Dayak. The Borneo tribal group which supplied many of the original trackers to the SAS for operations on the peninsula prior to the point in time at which the Orang Asli were won over by the government.

Emergency. The Malayan Emergency began in 1948 as a bid by the Malayan Communist Party to take control of the country. It was unilaterally declared over in 1960 but not formally concluded by treaty until 1989. The term Emergency was arrived at by the British to refer euphemistically to the war as a civil problem in an attempt to deny the communists legitimacy.

Gurkhas. Nepali troops long recruited by the British as mercenary fighters.

Iban. One of a dozen or so Borneo tribal groups. Many of these peoples were drafted into the Borneo Border Scouts.

Ipoh Tree. Sap from this plant, known by its botanical name as *antiaris toxicaria,* is boiled down into the purplish colored poison employed by Orang Asli hunters to kill their prey.

Jakun. One of three main Orang Asli tribal categories (generally referred to as Proto-Malay) who inhabit southern coastal areas and inland riverine territories.

Jampi. Orang Asli incantations.

Jarai. A Montagnard tribal group with Malayo-Polynesian qualities. Recruited by R.O.D. Noone to the anti-communist cause at the behest of GVN in Pleiku Province.

Kampong. Settlement or village.

Kuala Lumpur. The Malaysian capital city. The term means literally muddy estuary.

Malaya. For convenience the term "Malaya" is often employed in place of West Malaysia throughout this study and is meant to be exclusive of the Borneo territories. The word Malaysia came into being upon Malayan independence from Great Britain and is more contemporary in terms of the holistic nationalistic images which it evokes.

Mentri Besar. A Malay chief minister. Each state has one and there is also a Federal equivalent.

Merdeka. Independence from Great Britain. Achieved on August 31, 1957.

Mikong. A Sakai Rajah. Powerful individuals often of Orang Asli ancestry or mixed Malay/Orang Asli parentage that served as liaisons between peoples of the interior and the outside world. Usage of the term dates to Thai suzerainty over parts of Malaya.

Min Yuen. Peoples Movement. The ethnic Chinese based MCP/MRLA support organization replaced by the Asal Clubs.

Moi. The Vietnamese term of reference for the Montagnards, meaning savage. It parallels but does not quite approximate the pejorative designation "sakai" once employed derisively by some Malays to refer to the Orang Asli.

Montagnard. The generic term for highland peoples in Vietnam. The term translates from the French language as mountaineer.

Murut. North Borneo tribal peoples who formed the basis of the Sabah Border Scouts.

Orang Asli. A heterogenic aboriginal people of uncertain origin who are thought to be the first inhabitants of peninsular Malaysia. A literal translation of the term is "original people." Three broad overlapping categories of Malaysian aborigine are subsumed under the Orang Asli moniker: The Semang, Senoi, and Jakun—each possessing its own various subcomponents.

Parameswara. The founder of the Malacca Sultanate and first great Malay leader.

Peng Hulu. Headman, chief or district head. The title also has a connotation associated with leadership in war dating back to earlier times.

Proto-Malay. The current term of reference for half-dozen Orang Asli tribes once referred to as Jakun peoples. Use of this phraseology is somewhat akin to employment of the discarded term Saudara Lama. It suggests a strong Orang Asli/Malay relationship in which the Orang Asli are akin to the Malay.

Recce. Short for reconnaissance and pronounced reckee.

Saka. A tribal homeland or hereditary area within which related Orang Asli bands roam as hunter gatherer/slash and burn horticulturists.

Sakai. A derogatory term, connotative of slave or dependent, once employed by some Malays to refer to the Orang Asli.

Semai. A member of the Senoi tribal category.

Semang. One of three main Orang Asli tribal categories who inhabit parts of northern Malaysia and southern Thailand.

Senoi. A principal Orang Asli category comprising two further sub-groups—the Semai and Temiar. The Senoi reside in the mountain range which runs the length of central and north Malaysia.

Senoi Baruh. People of the plains. Orang Asli peoples of Senoi extraction living at lower elevations. Considered by some to be too "Kampungized" and out of touch with the jungle to make good Senoi Praaq troopers.

Senoi Jelmol. People of the hills. A generic term of reference employed by lowland Senoi (sic Senoi Baruh) for the deep jungle Orang Asli. Senoi Jelmol per se have traditionally made the best Senoi Praaq troopers.

Special Branch. A non uniform branch of the Royal Malaysia Police responsible for intelligence and political affairs.

Tanka. Necklace born charms worn by the Orang Asli to ward off harm.

Temiar. A sub-group of the Senoi tribe. Typically more aggressive than other Orang Asli.

Temuans. A land Jakun people living in riverain areas of Johore, Malacca, and Selangor. Referred to sometimes as Orang Hulu.

Tok Batin. Headman or chief who has a "letter of authority" with a more limited scope of responsibility than the traditional Peng Hulu.

Tuan. Traditional honorific title meaning lord or master in Bahasa. Currently used in contemporary Malay society as a polite and common form of address that translates as sir as opposed to lord or master.

Tunku. A term of reference for Malay royalty. It designates a prince.

Ulu. Literally means upriver or headwaters.

Vat 69. A hardcore commando operation and sometimes Senoi Praaq competitor known today as Commando 69.

Viet Cong. Later day Vietnamese communists (local to South Vietnam) and scourge of the Americans.

Viet Minh. Early Vietnamese communists and scourge of French forces.

Malaysian equivalent of the Rand Corporation. This think tank is the repository for Malaysian government thinking about regional security matters and a key source of GOM's foreign policy impetus. Dato Hassan, a veteran Malaysian national security counsel member and an astute observer of world politics, spent time with the author and helped to orient his thinking about regional security matters.

Jimoi, Mat bin. A Jakun and textbook example of the affirmative action-like benefits which acrue to the Senoi Praaq for rank and file members. Corporal Jimoi spoke about his life in the Senoi Praaq and shared information about the opportunities afforded to him and his family as a direct consequence of his service. In sum, a generation of Orang Asli entrepreneurs emerged over the past few years in many heretofore undeveloped parts of Malaysia as a direct consequence of Orang Asli participation in the Senoi Praaq. Upon severance retiring members of the Senoi Praaq typically return home at a relatively early age armed with a generous benefits package and a variety of skills. This provides them with the drive and ability to start up a variety of business ventures and helps the many remote Orang Asli villages which supply recruits to the Senoi Praaq to be better developed than they might otherwise have been without GOM sponsored infusions of talent and capital.

Jiwa, Mohamad. Director of the new JHEOA museum for the Orang Asli in Gombak at which there is an excellent exhibit devoted to the Senoi Praaq. The author was allowed to see the restricted materials in the collection not normally available to the public. Director Jiwa formerly controlled all firearms in the possession of the JHEOA and oversaw weapons and jungle survival training for new department members. In this respect the Senoi Praaq maintains strong ties with JHEOA and in fact cooperates closely with it in the interior. Symbiotic bonds between the two organizations and a genuine dependence upon one another for support builds strong personal relationships which administrative reorganizations across time do little to change.

Jones, Michael (Mike). An ex-regular Special Air Service (SAS) officer who served in Malaya during the Emergency and in Sarawak during Confrontation with Indonesia. Upon leaving the British Army he spent twenty years as a security consultant in Hong Kong and Southeast Asia, traveling widely in many areas in the region (inclusive of eight years in Thailand and its border areas). He maintains an interest in the Orang Asli and the Senoi Praaq developed during his early days in Malaya and is at present completing a study of Richard Noone and the influence exercised by this individual during a number of "Asian" conflicts, 1948-1974. Mr. Jones kindly read a draft of this book and helped the author correct a number of errors.

Khamis, Haji Ahmad bin. A career Senoi Praaq/JHEOA operative who served as Mohammed Ruslan bin Iskandar Abdullah's right hand at the Senoi Praaq and in other distinguished positions. A very careful individual who later became Director of Operations at the JHEOA and also served as the department's state director in Pahang. Haji Ahmed managed the Senoi Praaq air supply operation in Taiping, oversaw troop movements between competing obligations and is generally credited with maximizing Senoi Praaq performance in the field. The unit simply could not have survived without his input into day to day operations given the tremendous demands placed upon the organization during its progressive expansion over the years. Naturally, his comments and observations about the logistics behind Senoi Praaq operations provided invaluable insight into Senoi Praaq operational processes. Thus, not only was he indispensable to the organization but he was also of invaluable assistance in this research. His kindness towards the author helps guarantee the Senoi Praaq its proper recognition in Malaysia's history.

Leary, John D. A former member of the British Special Air Services (SAS) who cut his teeth in Malaya and author of a recent book, *Violence and the Dream People*, debunking the notion of some American anthropologists that the Orang Asli are by nature a peaceful people. Having lived the life, he and J. M. Gullick are unique among western authors writing about Malaya and/or the Orang Asli with the possible exceptions of a few other men like Spencer Chapman, R.O.D. Noone, J. P. Cross and Herbert Andrew. Leary provided a great deal of insight into the more macabre aspects of the Orang Asli psyche, not to mention their capacity for violence. Leary is very much a realist and his assessment of the Orang Asli is possessed by great lucidity.

Lee, Chew Jang. A Senoi (Semoq Beri sub group) with a Chinese name and career employee at the JHEOA. This individual's family members suffered immeasurably during the Emergency yet they were fortunate to have had a positive experience with loyalist Chinese. Hence, the adoption in better times of multiple Chinese names as a gesture of their appreciation. He spoke with the author at some length about the extent to which the Emergency ravaged the various Orang Asli population groups. Information about his exposure to this rather extensive, sad and all too recent human tragedy helped to keep the research in perspective; otherwise, the author's penchant for adventuresome, sometimes heraldic, pomp would tend to overshadow local sensitivities induced by the fighting, obscure the overall picture and deprive readers of an objective grasp of those dark days.

Ley, Charles H. By far the most successful squadron commander every to have served in the Senoi Praaq. Mr. Ley journeyed to Malaya in February 1952 as a rubber planter in Kuantan and acquired extensive local knowledge prior to joining the unit. He also accompanied R.O.D. Noone to Vietnam, fought on Borneo and eventually commanded the Sabah Border Scouts. Mr. Ley then took over R.O.D. Noone's SEATO appointment in Thailand, for a period of three years, upon the later's 1973 death in Bangkok. Subsequently he became Counselor for Hongkong Affairs in the British Embassy, Bangkok over a six year period dealing with illicit narcotic warfare. As such Mr. Ley was the U.K. and Hong Kong representative to the Foreign Anti-Narcotic Committee (FANC). The author is grateful to Mr. Ley for sharing some of his experiences and for trouble shooting the manuscript.

Mathews, Phillip. Associate Director of ISIS Malaysia. A congenial fellow who made the author welcome at ISIS who is best described as the consummate public relations specialist. Director Mathews gave generously of his time, put the resources of ISIS at the author's disposal and personally made arrangements to facilitate his research.

Mustapha, Jamshah bin. A liaison officer at the Royal Malaysia Police. Jamshah worked tirelessly and put in a lot of time to ensure that meetings in Kuala Lumpur and in field camps with various Senoi Praaq and other police personalities went smoothly and were productive.

Navaratnam, A. Founder of the Vat 69 and known within the Malaysian special forces community as Nava. Author of the soon to be published history of the Vat 69, *The Spear and the Kerambit.* Information provided by this intense individual enabled this author to contrast the Senoi Praaq and the Vat 69 with other Malaysian paramilitary forces and thus to better understand their respective roles relative to the conventional military establishment.

Noor, Mohamed bin. A Semai referred to with some amusement and for good reason as 007 by Senoi Praaq fighters. ASP Noor matriculated into the Senoi Praaq as 07 and is in fact the last living member of the initial ten Orang Asli recruited by the SAS. He was also active in the Asal movement, prior to his capture by Charles Ley along the Cameron Highlands road, and is the quintessential guerilla fighter. A bouncy and energetic legend among the Senoi Praaq renowned in later years for his ability to summon helicopters out of thin air. ASP Noor recently retired as a Police Gerakan Am (PGA) battalion commander with the rank of Assistant Superintendent of Police and holds the highest rank achieved by any Orang Asli in the PGA.

Ruslan, (Mohamed Ruslan bin Iskandar Abdullah). Born Edmund Colin Ritson Dewsnup, British expatriate, adventurer and last Anglo commander of the Senoi Praaq. A charming yet impassive host who personifies the classic and impersonal Chinese warlord of yore. Indeed a great irony given his own well-earned and painful dislike for some of the more obstinate Han peoples to be found in Malaysia. One of the few brave men sent by Great Britain to fight communism in Malaya who survived the many ordeals imposed by the Emergency upon those who fought all its battles. No other British subject is known to have experienced the entire duration of the war and lived long enough thereafter to tell the tale. Men like this made the British Empire great, arrested Malayan communism in its tracks, and ultimately through their stalwart sacrifices helped to stem the tide against the Red threat in Southeast Asia.

Saban, Udah. A Semai Sergeant Major assigned to the Bidor camp. Sergeant Saban shared his insights into the art of successful blowgun use and elaborated upon the Orang Asli belief in magic. Charms and incantations are part of an ancient form of Orang Asli magic through which injury and or accident may be warded off. Charms per se are often reinforced by a secret mantra of two words and insure the wearer is immune to gunfire. The logic behind these charms is circular in that they only work if one's faith in the underlying magic is strong enough to guarantee effectivity. From Udah's demeanor one can easily perceive how the Senoi Praaq rank and file clearly have faith in their Sergeants, if not protective talismans.

Saharudin, Arbi bin. Director for the JHEOA office of Security Affairs in Gombak and part of a second Orang Asli levy recruited into the original Senoi Praaq. Director Saharudin is thus among the first 30 Orang Asli to join the unit and did so in early 1957 at the age of 14. He became the unit's youngest member and first saw action in August of that same year. This date coincided with Malaya's independence from Great Britain but the country was by no means secure. The many MRLA fighters who remained in the deep jungle actively sought out the Senoi Praaq and made several concerted attempts to annihilate the fledgling unit. Daily contacts with MRLA guerillas occurred and Director Saharudin considers himself fortunate indeed to have survived. Nonetheless he went on to serve two extended tours with the unit and make a career out of jungle fighting prior to joining the JHEOA. This individual captured the author's imagination and provided the initial encouragement for a book focusing on the Senoi Praaq. Arbi Saharudin is Semai.

Thambpillay, R. Retired Superintendent of Police, active in community service in the Ipoh area of Perak State, an author of *God's Little Acre*, a commemorative book published on the 50th anniversary of the Malayan Emergency. He saw extensive action in twelve major jungle operations in the Betong Salient. Supt. Thambpillay conveyed to the author his deep appreciation and understanding of the depth of the suffering the fighting brought not only to the Orang Asli but also to the Malaysian and Commonwealth nationals involved in the fighting. After the excitement of victory fades the sadness of human tragedy remains.

Yusof, Udah bin. A Jakun and presently the Tok Batin at a large Orang Asli settlement beyond Gombak. A rare individual in the league of Mohamed bin Noor and Arbi bin Saharudin with extensive combat experience during both the Emergency and the Konfrontasi. A large man for an Orang Asli (much like his older brother—Bah Tangga, R.O.D. Noone's fearsome bodyguard) who shared grass roots information about the Senoi Praaq. Most definitely someone that it would not be a good idea to meet in the dark.

UNPUBLISHED DOCUMENT SOURCES

Arkib Negara, Kuala Lumpur
John Fitzgerald Kennedy Library, Boston, Massachusetts
National Archives and Records Administration, College Park, Maryland
Public Record Office, Kew
Royal Malaysia Police, Kuala Lumpur

PUBLISHED DOCUMENTS AND UNPUBLISHED PAPERS

Royal Malaysia Police, U.S. State Department and British Ministry of Defence

Department of State, *Foreign Relations of the United States, 1961-1963*, Volume 3, Vietnam, January-August 1963 (Washington, D.C.: GPO, 1991).

________. *Foreign Relations of the United States, 1964-1968*, Volume 1, Vietnam, 1964 (Washington, D.C.: G PO, 1992).

________. *Foreign Relations of the United States, 1961-1963*, Volume 1, Vietnam, 1961 (Washington, D.C.: GPO, 1988).

Ministry of Defense. Briggs, Sir Harold, Lieut. General, K.C.I.E., C.B., C.B.E., D.S.O., Director of Operations. *Report on the Emergency in Malaya from April 1950 to November 1951*. (15 February 1953): 1-53. Kuala Lumpur: Unpublished Paper.

Royal Malaysia Police. *Table of Organization and Summary, The Royal Malaysia Police*. 1998.

________. Mustapha, Jamshah, DSP. *The Royal Malaysia Police: Organization and Background*. Bukit Aman, 1999.

Alsagoff, Syed Zainal Abideen, P.P., K.M.N., P.T.S., P.P.T, A.M.N., *Senoi Praaq: Pegawai Pemerintah. Batalion KE 20 Pasukan Polis Hutan Bidor, Perak*. (30 July 1990): 1-83. Unpublished Paper.

Zain, bin Mohamed Anuar, DSP. *Roles of the Senoi Praaq Towards Defense and Development of the Country*. (Kuala Lumpur: August 1989): 1-46. Malaysian Armed Forces Staff College 1989. Unpublished Paper.

Books, Scholarly Journals and Unpublished Papers

Andrew, Herbert. *Who Won the Malayan Emergency?* Singapore: Graham Brash, 1995.

Aslie, Haji Mohamed Reduan and Haji Mohamed Radzuan Ibrahim. *Polis Diraja Malaysia: Sejarah, Peranan Dan Cabaran*. Kuala Lumpur: Kumpulan Karangkraf, 1984.

Bailey, Thomas and Bruce Lawlor. "The Phoenix." In *Everything We Had: An Oral History of the Vietnam War by Thirty-three American Soldiers Who Fought It*, Edited by Albert Santoli, 199-202. New York: Random House, 1981.

Barber, Noel. *The War of the Running Dogs. How Malaya Defeated the Communist Guerrillas, 1948-1960*. London: Collins, 1971.

Beamon, Mike. "The Green-Faced Frogmen." In *Everything We Had: An Oral History of the Vietnam War by Thirty-three American Soldiers Who Fought It*, Edited by Albert Santoli, 203-219. New York: Random House, 1981.

Bergot, Erwin. *Les Heros Oublies*. Paris: Bernard Grasset, 1975.

Bernard, Edward. *Going Home*. Philadelphia: Dorrance & Company, 1973.

Bertrand, Gabrielle. *Le Peuple de la Jungle: Hommes, Betes et Legendes du Pays Moi*. Paris: Editions "Je Sers", 1955.

Bierman, John and Colin Smith. *Fire in the Night: Wingate of Burma, Ethiopia, and Zion*. New York: Random House, 1999.

Bodard, Lucien. *La Guerre D'indochine: L'aventure*. Paris: Gallimard, 1967.

Bodard, Lucien. *The Quicksand War: Prelude to Vietnam*, Translated and with an

introduction by Patrick O'Brian. Boston: Little, Brown and Company, 1967.
Bonnecarrere, Paul. *Par le sang verse: la Legion Etrangere en Indochine*. Paris: Artheme Fayard, 1969.
Bui, Diem with David Chanoff. *In the Jaws of History*. Boston: Houghton Mifflin Company, 1987.
Burchett, Wilfred G. *The Furtive War: The United States in Vietnam and Laos*. New York: International Publishers Co., 1963.
Burchett, Wilfred G. *Vietnam: Inside Story of the Guerilla War*. New York: International Publishers Co., 1965.
Carey, Iskandar. Orang Asli: The Aboriginal Tribes of Peninsular Malaysia. Kuala Lumpur: Oxford University Press, 1976.
Chapman, Spencer F. The Jungle Is Neutral. London: Catto and Windus, 1947.
Colby, William with James McCargor. *Lost Victory: A Firsthand Account of America's Sixteen Year Involvement in Vietnam*. Chicago: Contemporary Books Inc., 1989.
Collins, James Lawton Jr. *The Development and Training of the South Vietnamese Army, 1950-1972*. Washington: Department of the Army, 1975.
Couillard, Marie-Andree. "Les Malais et les Sakai: Quelques Reflections sur les Rapports Sociaux dans la Peninsule Malaise." *Anthropologies et Societes* 7, no. 2 (1983): 91-109.
Clutterbuck, Richard. *The Long Long War: The Emergency in Malaya 1948-1960*, with a forward by Sir Robert Thompson. London: Cassell, 1967.
Crawford, Steve. *The SAS Encyclopedia: The Definitive Companion to the World's Crack Regiment*. Miami: Lewis International Inc., 1998.
Cross, John. *Jungle Warfare: Experiences and Encounters*. London: Arms and Armour Press, 1989.
________. *A Face Like A Chicken's Backside: An Unconventional Soldier in South East Asia*, 1948-1971. London: Greenhill Books, 1996.
Davies, Barry. *SAS: The Illustrated History*. London: Virgin Books, 1997.
Dickens, Peter. *SAS: Secret War in South-East Asia*. New York: Ivy Books, 1995.
Fall, Bernard B. *Last Reflections on a War*. Garden City: Doubleday, 1967.
Geraghty, Tony. *Who Dares Wins: The Story of the Special Air Service 1950-1982*. London: Fontana Collins, 1983.
Gia, Tac and Adkins, E.H: *A Study of Montagnard Names in Vietnam: Khao Cu'u Ve Ten Ngu'oi Thu'o'ng Tai Viet-Nam*. Saigon: Michigan State University Vietnam Advisory Group, 1962.
Gullick, J.M. *Indigenous Political Systems of Western Malaya*. London: Athlone Press, 1988.
Hackworth, David H. *About Face: The Odyssey of an American Warrior*. New York: Simon and Schuster, 1989.
Halberstam, David. *The Best and the Brightest*. Greenwich, Connecticut: Fawcett Publications, Inc., 1972.
Hamilton, Donald W. *The Art of Insurgency: American Military Policy and the Failure of Strategy in Southeast Asia*, with a foreword by Cecil B. Currey.

Westport, Connecticut: Praeger, 1998.
Hannah, Norman B., *The Key To Failure.* Lanham, MD.: Madison Books, 1987.
Hickey, Jerald Cannon. *Free in the Forest: Ethnology of the Vietnamese Highlands 1954-1976.* New Haven: Yale University Press, 1982.
Hill, A. H. et al. *The Keris and Other Malay Weapons.* London: the Malayan Branch of the Royal Asiatic Society, 1998.
Holman, Dennis. Noone of the Ulu, With a Foreword by Field-Marshall Sir Gerald Templer, G.C.B., G.C.M.G., K.B.E., D.S.O. London: Heinemann, 1958.
Hyde, Douglas A. *Confrontation in the East: A Background Book.* Chester Springs, PA: Dufour Editions, 1965.
James, Harold and Dennis Sheil-Small. *The Undeclared War: The Story of the Indonesian Confrontation 1962-1966.* Totowa, NJ: Rowman and Littlefield, 1971.
Jackson, Robert. *The Malayan Emergency: The Commonwealth's Wars 1948-1960.* London: Routledge, 1991.
Jeshurun, Chandran. *Malaysian Defense Policy: A Study in Parliamentary Attitudes 1963-1973.* Kuala Lumpur: University of Malaya Press, 1980.
Joiner, Charles A. *The Politics of Massacre: Political Processes in South Vietnam.* Philadelphia: Temple University Press, 1974.
Jumper, Roy. "The Communist Challenge in South Vietnam," *Far Eastern Survey* XXV (November 1956) 161-168.
________. "Problems of Public Administration in South Vietnam," *Far Eastern Survey* 16 (December 1957): 183-190.
________. "The Cao Dai of Tay Ninh: The Politics of a Political-Religious Sect in South Vietnam." In *Asian Studies,* ed. Balkrishna G. Gokhale, 142-154. Bombay: Popular Prakashan, 1966.
Jumper, Roy Davis Linville. *Power and Politics: The Story of Malaysia's Orang Asli.* Lanham, MD: University Press of America, 1997.
________. *Orang Asli Now: The Orang Asli in the Malaysian Political World.* Lanham, Maryland: University Press of America, 1999.
________. "Malaysia's Senoi Praaq Special Forces," *International Journal of Intelligence and Counter Intelligence* 13 (Spring 2000): 64-93.
Kahin, George McTurnan and John W. Lewis. *The United States in Vietnam.* New York: The Dial Press, 1967.
Kaiser, David. *American Tragedy: Kennedy, Johnson, and the Origins of the Vietnam War.* Cambridge: Harvard University Press, 2000.
Karnow, Stanley. *Vietnam: A History.* New York: The Viking Press, 1983.
Kelly, Francis J. *U.S. Army Special Forces, 1961-1971.* Washington: Department of the Army, 1973.
Kiem, Thai Van. *Viet-Nam: d'hier et d'aujourd'hui.* Saigon: Ouvrage publie sous le patronage du Departement de l'Education Nationale du Viet-Nam et de la Commission Nationale du Viet-Nam pour L'UNESCO, 1956.
LaPalombara, Joseph, ed. *Bureaucracy and Political Development.* Princeton, New Jersey: Princeton University Press, 1963.

Leary, John. *Violence and the Dream People: The Orang Asli in the Malayan emergency, 1948-1960*. Athens, OH: Ohio University Press, 1995.
Mackie, J.A.C. *Konfrontasi: The Indonesian-Malaysia Dispute, 1963-1966*. Kuala Lumpur: Oxford University Press, 1974.
Maclean, Fitzroy. *Eastern Approaches*. London: Transworld Publishers, Ltd., 1975.
Masterman, J. C., *The Double-Cross System*. New York: Avon Books, 1972.
McNamara, Robert S. with Brian VanDeMark. *In Retrospect: The Tragic Lessons of Vietnam*. New York: Times Books, 1995.
Miller, Harry. *Jungle War in Malaya: The Campaign Against Communism, 1948-1960*. London: Arthur Barker Limited, 1972.
Nolting, Frederick. *From Trust to Tragedy: The Political Memoirs of Frederick Nolting, Kennedy's Ambassador to Diem's Vietnam*. New York: Praeger, 1988.
Noone, Richard O.D. and Dennis Holman. In *Search of the Dream People*, with a foreword by Field-Marshal Sir Gerald Templer. New York: William Morrow and Company, 1972.
O'Ballance, Edgar. *Malaya: The Communist Insurgent War, 1948-1960* . London: Faber and Faber Limited, 1966.
Oberdorfer, Don. *Tet*. New York: Avon, 1972.
Pagniez, Yvonne. *Francais D'Indochine: l'Aventure Vecue*. Paris: Editions Flammarion, 1953.
Parkinson, Northcote C. *Templer in Malaya*. Singapore: Donald Moore Ltd., 1954.
Prados, John. *President's Secret Wars: CIA and Pentagon Covert Operations from World War II through Iranscam*. New York: William Morrow, 1986.
Prados, John. *The Hidden History of the Vietnam War*. Chicago: Ivan R. Dee, 1995.
Prados, John. *The Blood Road: The Ho Chi Minh Trail and the Vietnam War*. New York: John Wiley & Sons, Inc., 1999.
Rahim, Lily Zubaidah. "Singapore-Malaysia Relations: Deep-seated Tensions and Self-fulfilling Prophecies." *Journal of Contemporary Asia* 29 (1999): 38-56.
Rambo, Terry A. "Why are Semang? Ecology and Ethnogenisis of Aboriginal Groups in Peninsular Malaysia." In *Ethnic Diversity and the Control of Natural Resources in Southeast Asia*, Edited by A. Terry Rambo, Kathleen Gillogly, and Karl L. Hutterer, 19-36. Ann Arbor: University of Michigan Press, 1988.
Rose, George. *The British in Vietnam: How the Twenty-Five-Year War Began*. London: Panther Books Limited, 1970.
Sheehan, Neil, et al. *The Pentagon Papers: as Published by the New York Times*. New York: Quadrangel Books, 1971.
Short, Anthony. *The Communist Insurrection in Malaya, 1949-1960*. New York: Crane, Russak, 1975.
Short, Anthony. *The Origins of the Vietnam War*. London: Longman, 1989.
Shultz, Richard H. *The Secret War Against Hanoi: Kennedy's and Johnson's Use*

of Spies, Saboteurs and Covert Warriors in North Vietnam. New York: Harper Collins Publishers Inc., 1999.

Slimming, John. *Temiar Jungle: A Malayan Journey*. London: John Murray, 1958.

Stacey, Tom. *The Hostile Sun*. London: Gerald Duckworth & Co. LTD., 1954.

Tang, Truong Nhu. *Journal of a Vietcong*. London: Jonathan Cape, 1986.

Thompson, Sir Robert. *Peace Is Not At Hand*. London: Chatto & Windus, 1974.

Thompson, Sir Robert. *Defeating Communist Insurgency: The Lessons of Malaya and Vietnam*. New York: Praeger, 1970.

Tourison. Sedgwick. *Secret Army Secret War: Washington's Tragic Spy Operation in North Vietnam*, Annapolis. Naval Institute Press, 1995.

Valentine, Douglas, *The Phoenix Program*. New York: William Morrow and Company, Inc., 1990.

Warren, William. *The Legendary American: The Remarkable Career and Strange Disappearance of Jim Thompson*. Boston: Houghton Mifflin Company, 1970.

Webster, Donovan. *Aftermath: The Remnants of War*. New York: Pantheon Books, 1996.

Westmoreland, General William C. *A Soldier Reports*. New York: Doubleday & Company, Inc., 1976.

Wilkinson, R. J., ed., *Papers on Malay Subjects*. London: Oxford University Press, 1971.

Wolff, Leon. *In Flanders Fields*. Greenwood, Westport, CT: 1984.

Newspapers and Non-Academic Journals

"3,300 Illegal Aliens Held in 2 Months." *New Straits Times*, 13 May 1995, 23.

Abdullah, Nik Imran. "Police to Help Curb Cross-Border Crimes." *New Straits Times*, 26 February 1998, 12.

"Agencies Must Act in Concert." *New Straits Times*, 2 March 1998, 12.

Andres, Leslie, et al. "Pos Dipang Survivors Rebuilding their Lives." *New Straits Times*, 4 September 1996, 10.

"Bringing First People into the Fold." *New Straits Times*, 9 August 1997, 1.

"Communist Terrorist Activities Contained." *BBC Summary of World Broadcasts*, 12 June 1985, (Part 3 The Far East; B. Internal Affairs; Malaysia; FE/7975/B/1).

Emmanuel, Tony, et al. "Riot Police Seize Weapons Stockpile from Camp." *New Straits Times,* 11 June 1995, 1.

"From Fighting Reds to General Police Work." *New Straits Times*, 3 February 1995, 20.

Getlin, Josh. "Celebrates First Thanksgiving; Montagnard Refugees Find New Life in U.S."*Los Angles Times*, 30 November 1986, 1.

Golden, Tim. "CIA Error Sent Guns to Rebels in Colombia," *Herald Tribune*, November 7, 2000.

"Government Urged to Review Role of Arms-Bearing Village Militia." *The Straits Times*, 28 March 1995, 19.

Jayasankaran, S. "Under the Gun." *Far Eastern Economic Review* 161 (March 1998): 20.

Joheng, Jalina. "General Duty Police to Help Nab Illegals." *New Straits Times*, 1 March 1998, 1.

Jones, Mike. "Chen Peng and Mangoes." *Australian Veterans Association (Malaysia and Borneo) Newsletter*, August 1998, 20-21.

Karniol, Robert and Joris Janssen Lok. "Caution is the Catchword in Drive for Defense Industrialization." *Jane's Defense Weekly* 28 (November 1997): 38.

"K. L. Steps Up Checks to Prevent Inflow of Illegal Immigrants." *The Straits Times*, 11 February 1998.

"Malaysia to Upgrade Border Facilities to Prevent Illegal Entry, Smuggling." *The Xinhua News Agency*, 15 (February 1997): item no. 0215161.

Mahsinah, Sharifah. "43 Held in Big Contraband Ops by Police" *New Straits Times*, 22 December, 1995, 5.

"Mohamed: Senoi Praaq Team Also Member of SMART." *New Straits Times*, 3 September 1996, 5.

Ng, Freddie. "IGP to Chair Meeting on Illegal Immigrants." *New Straits Times*, 10 March 1998, 8.

"PFF Personnel Shot at During Border Patrol." *New Straits Times*, 29 May 1997, 3.

"Police on Alert at Border to Check Entry of Illegals." *New Straits Times*, 7 March 1996, 16.

Rahman, Azlyn A. "Esso to Equip Senoi Praaq Library with PC Facilities." *New Straits Times*, 17 September 1998, 8.

"Ramli: Search to Go On Until All Are Found." *New Straits Times*, 4 September 1996, 10.

Singh, Sarban. "80,000 Members in Terriers by 2000." *New Straits Times*, 28 April 1997, 10.

"Six Schoolboys Lost in Jungle Rescued." *New Straits Times*, 15 April 1997, 19.

"Special Police Units to Help Customs Check Traders' Import Permits." *The Straits Times*, 25 May 1992, 19.

"Stepping up Campaign to Nab Illegals." *New Straits Times*, 6 July 1997, 11.

Teo, John. "U.S. Must Not Try to Enforce Democracy on the Rest of Us." *New Straits Times*, 25 April 2000, 15.

"Territorial Army Recruits 200 Orang Asli in Pahang." *New Straits Times*, 15 May 1996, 8.

Index

About the Author

ROY DAVIS LINVILLE JUMPER is a political scientist who is trained in the fields of Comparative Politics, International Relations, and Public Policy. He has published throughout his academic career, holds two advanced degrees, works in the U.S. Navy PACE program, and is a Francophile. His personal experiences and interests include extensive travel in Europe, Asia, and the Middle East where he encountered different belief systems and always sought to promote a positive image of the United States. He is the author of two previous books: *Orang Asli Now* (1999) and *Power and Politics* (1997).

CPSIA information can be obtained
at www.ICGtesting.com
Printed in the USA
BVHW042104040320
574144BV00011B/181

9 780313 315152